THERMAL ANALYSIS

THERMAL ANALYSIS

FUNDAMENTALS AND APPLICATIONS TO POLYMER SCIENCE

T. Hatakeyama and F.X. Quinn

National Institute of Materials and Chemical Research, Ibaraki, Japan

JOHN WILEY & SONS

Chichester · New York · Brisbane · Toronto · Singapore

ACO 2295

QD
79
.T38
H38
1994

Library of Congress Cataloging-in-Publication Data

Hatakeyama, T.
   Thermal analysis : fundamentals and applications to polymer
   Science / T. Hatakeyama and F.X. Quinn.
     p.   cm.
   Includes bibliographical references and index.
   ISBN 0–471–95103–X
   1. Thermal analysis. I. Quinn, F.X. II. Title.
   QD79.T38H38   1994   94–4641
   543′.086—dc20   CIP

British Library Cataloguing in Publication Data
A catalogue record for this book is available from the British Library

ISBN 0 471 95103 X

Typeset in 10/12pt Times by Alden Multimedia, Northampton.
Printed and bound in Great Britain by Biddles Ltd, Guildford, Surrey.

CONTENTS

PREFACE

We are grateful to those many friends, M. Maezono, Z. Liu, K. Nakamura, T. Hashimoto, S. Hirose, H. Yoshida and C. Langham, whose considerable input helped us to write this book.

We would like to extend special thanks to Hyoe Hatakeyama, who made many valuable suggestions when authors from different backgrounds encountered various problems. Without his encouragement this book could not have been written.

The book is the result of the merging of ideas from both East and West. We hope that readers will find it useful in their work. As Confucius said, it is enjoyable when friends come from far places and work together for the same purpose.

Tsukuba   T.H.

June, 1994   F.X.Q.

1 THERMAL ANALYSIS

1.1 Definition

The term thermal analysis (TA) is frequently used to describe analytical experimental techniques which investigate the behaviour of a sample as a function of temperature. This definition is too broad to be of practical use. In this book, TA refers to conventional TA techniques such as differential scanning calorimetry (DSC), differential thermal analysis (DTA), thermo-gravimetry (TG), thermomechanical analysis (TMA) and dynamic mechanical analysis (DMA). A selection of representative TA curves is presented in Figure 1.1.

TA, in its various guises, is widely employed in both scientific and industrial domains. The ability of these techniques to characterize, quantitatively and

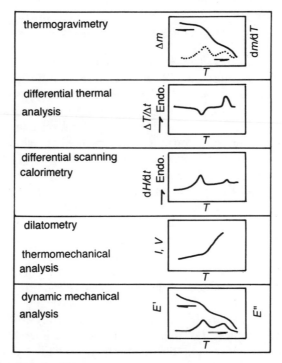

Figure 1.1. Representative TA curves

qualitatively, a huge variety of materials over a considerable temperature range has been pivotal in their acceptance as analytical techniques. Under normal conditions only limited training of personnel is required to operate a TA instrument. This, coupled with the fact that results can be obtained relatively quickly, means that TA is employed in an ever increasing range of applications. However, the operational simplicity of TA instruments belies the subtlety of techniques which, if improperly practised, can give rise to misleading or erroneous results. The abundance of results of dubious integrity in both the academic literature and industrial performance reports underlines the extent and seriousness of this problem.

1.2 Characteristics of Thermal Analysis

The advantages of TA over other analytical methods can be summarized as follows: (i) the sample can be studied over a wide temperature range using various temperature programmes; (ii) almost any physical form of sample (solid, liquid or gel) can be accommodated using a variety of sample vessels or attachments; (iii) a small amount of sample ($0.1\,\mu g$–$10\,mg$) is required; (iv) the atmosphere in the vicinity of the sample can be standardized; (v) the time required to complete an experiment ranges from several minutes to several hours; and (vi) TA instruments are reasonably priced. In polymer science, preliminary investigation of the sample transition temperatures and decomposition characteristics is routinely performed using TA before spectroscopic analysis is begun.

TA data are indirect and must be collated with results from spectrscopic measurements [for example, NMR, Fourier transform infrared (FTIR) spectroscopy, X-ray diffractometry] before the molecular processes responsable for the observed behaviour can be elucidated. Irrespective of the rate of temperature change, a sample studied using a TA instrument is measured under non-equilibrium conditions, and the observed transition temperature is not the equilibrium transition temperature. The recorded data are influenced by experimental parameters, such as the sample dimensions and mass, the heating/cooling rate, the nature and composition of the atmosphere in the region of the sample and the thermal and mechanical history of the sample. The precise sample temperature is unknown during a TA experiment because the thermocouple which measures the sample temperature is rarely in direct contact with the sample. Even when in direct contact with the sample, the thermocouple cannot measure the magnitude of the thermal gradients in the sample, which are determined by the experimental conditions and the instrument design. The sensitivity and precision of TA instruments to the physicochemical changes occuring in the sample are relatively low compared with spectroscopic techniques. TA is not a passive experimental method as the high-order structure of a sample (for example, crystallinity, network formation, morphology) may change during the measurement. On the other hand, samples can be annealed, aged, cured or have their previous thermal history erased using these instruments.

1.3 Conformation of Thermal Analysis Instruments

The general conformation of TA apparatus, consisting of a physical property sensor, a controlled-atmosphere furnace, a temperature programmer and a recording device, is illustrated in Figure 1.2. Table 1.1 lists the most common forms of TA. Modern TA apparatus is generally interfaced to a computer (work station) which oversees operation of the instrument controlling the temperature range, heating and cooling rate, flow of purge gas and data accumulation and storage. Various types of data analysis can be performed by the computer. A trend in modern TA is to use a single work station to operate several instruments simultaneously (Figure 1.3).

TA apparatus without computers is also used where the analogue output signal is plotted using a chart recorder. Data are accumulated on chart paper and calculations performed manually. The quality of the data obtained is not diminished in any way. The accuracy of the results is the same provided that the apparatus is used properly and the data are analysed correctly. Some instruments are equipped with both a computer and a chart recorder in order, for example, to evaluate ambiguous shifts in the sample baseline.

1.4 Book Outline

This text is designed to acquaint and orientate newcomers with TA by providing a concise introduction to the basic principles of instrument operation, advice on sample preparation and optimization of operating conditions and a guide to interpreting results. The text deals with DSC and DTA in Chapters 2 and 3. TG is described in Chapter 4. In Chapter 5 the application of these TA techniques to polymer science is presented. Other TA techniques

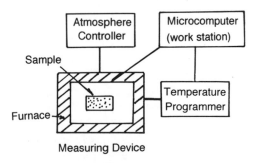

Figure 1.2. Block diagram of TA instrument

Table 1.1. Conventional forms of TA

Property	TA method	Abbreviation
Mass	Thermogravimetry	TG
Difference temperature	Differential thermal analysis	DTA
Alternating temperature	Alternating current calorimetry	ACC
Enthalpy	Differential scanning calorimetry	DSC
Length, volume	Dilatometry	
Deformation	Thermomechanical analysis	TMA
	Dynamic mechanical analysis	DMA
Electric current	Thermostimulated current	TSC
Luminescence	Thermoluminescence	TL

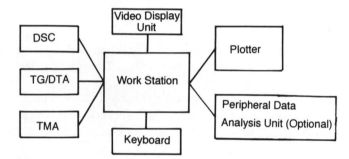

Figure 1.3. Simultaneous operation of several TA instruments using a central work station

are briefly described for completeness in Chapter 6. The Appendices include a glossary of TA terms, a survey of standard reference materials and TA conversion tables.

Although primarily pitched at newcomers, this book is also intended as a convenient reference guide for more experienced users and to provide a source of useful TA information for professional thermal analysts.

2 DIFFERENTIAL THERMAL ANALYSIS AND DIFFERENTIAL SCANNING CALORIMETRY

2.1 Differential Thermal Analysis (DTA)

The structure of a classical differential thermal analyser is illustrated in Figure 2.1. The sample holder assembly is placed in the centre of the furnace. One holder is filled with the sample and the other with an inert reference material, such as α-alumina. The term 'reference material' used in TA is frequently confused with the term 'standard reference material' used for calibration, since in many other analytical techniques the same material is used for both purposes. However, a reference material in TA is a thermally inert substance which exhibits no phase change over the temperature range of the experiment. Thermocouples inserted in each holder measure the temperature difference between the sample and the reference as the temperature of the furnace is controlled by a temperature programmer. The temperature ranges and compositions of

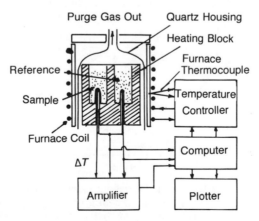

Figure 2.1. Schematic diagram of classical DTA apparatus

Table 2.1. Thermocouples commonly used in DTA

Thermocouple	Electric terminal		Recommended operating range/K
	+	−	
Cu–Constantan[a]	Cu	Constantan	90–600
Chromel[b]–Constantan	Chromel	Constantan	90–1000
Chromel–Alumel[c]	Chromel	Alumel	270–1300
Pt–Pt/Rh[d]	Pt/Rh	Pt	500–1700

[a]Constantan: Cu 60–45%, Ni 40–55%.
[b]Chromel: Ni 89%, Cr 9.8%, trace amounts of Mn and Si_2O_3
[c]Alumel: Ni 94%, Mn 3%, Al 2%, Si_2O_3 1%.
[d]Pt/Rh: Pt 90%, Rh 10%.

commonly used thermocouples are listed in Table 2.1. The thermocouple signal is of the order of millivolts.

When the sample holder assembly is heated at a programmed rate, the temperatures of both the sample and the reference material increase uniformly. The furnace temperature is recorded as a function of time. If the sample undergoes a phase change, energy is absorbed or emitted, and a temperature difference between the sample and the reference (ΔT) is detected. The minimum temperature difference which can be measured by DTA is 0.01 K.

A DTA curve plots the temperature difference as a function of temperature (scanning mode) or time (isothermal mode). During a phase transition the programmed temperature ramp cannot be maintained owing to heat absorption or emission by the sample. This situation is illustrated in Figure 2.2, where the

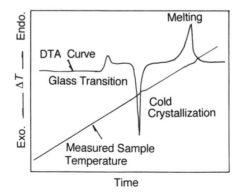

Figure 2.2. Schematic illustration of the measured sample temperature as a function of time for a polymer subjected to a linear heating ramp, and the corresponding DTA curve. In the region of the phase transitions the programmed and measured sample temperatures deviate significantly

temperature of the sample holder increases above the programmed value during crystallization owing to the exothermic heat of crystallization. In contrast, during melting the temperature of the sample holder does not increase in response to the temperature programmer because heat flows from the sample holder to the sample. Therefore, the true temperature scanning rate of the sample is not constant over the entire temperature range of the experiment.

Temperature calibration is achieved using standard reference materials whose transition temperatures are well characterized (Appendices 2.1 and 2.2), and in the same temperature range as the transition in the sample. The transition temperature can be determined by DTA, but the enthalpy of transition is difficult to measure because of non-uniform temperature gradients in the sample due to the structure of the sample holder, which are difficult to quantify. This type of DTA instrument is rarely used as an independent apparatus and is generally coupled to another analytical instrument for simultaneous measurement of the phase transitions of metals and inorganic substances at temperatures greater than 1300 K.

2.1.1 CUSTOM DTA

Many DTA instruments are constructed by individual researchers to carry out experiments under specialized conditions, such as high pressure and/or high temperature. Figure 2.3 shows an example of a high-pressure custom DTA

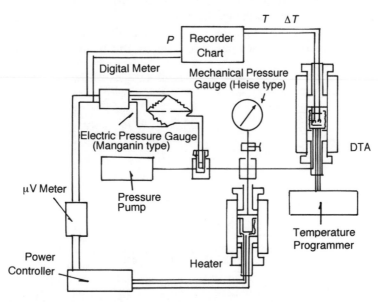

Figure 2.3. Schematic diagram of a high-pressure custom DTA apparatus (courtesy of Y. Maeda)

instrument. The pressure medium used is dimethylsilicone (up to 600 MPa) or kerosene (up to 1000 MPa) and the pressure is increased using a mechanical or an electrical pump. The temperature range is 230–670 K at a heating rate of 1–5 K/min. The pressure range of commercially available high-pressure DTA systems is 1–10 MPa to guarantee the stability and safety of the apparatus. In this case, the excess pressure is generated using a purge gas (CO_2, N_2, O_2). The DTA heating curves of polyethylene measured over a range of pressures are presented in Figure 2.4.

DTA systems capable of measuring large amounts of sample (> 100 g) have been constructed to analyse inhomogeneous samples such as refuse, agricultural products, biowaste and composites.

2.2 Quantitative DTA (Heat-flux DSC)

The term heat-flux differential scanning calorimeter is widely used by manufacturers to describe commercial quantitative DTA instruments. In quantitative DTA, the temperature difference between the sample and reference is measured as a function of temperature or time, under controlled temperature conditions. The temperature difference is proportional to the change in the heat flux (energy input per unit time). The structure of a quantitative DTA system is shown in Figure 2.5. The conformation of the sample holder assembly is different from that in a classical DTA set-up. The thermocouples are attached to the base of the sample and reference holders. A second series of thermocouples measures the temperature of the furnace and of the heat-sensitive plate. During a phase change heat is absorbed or emitted by the sample, altering the heat flux through the heat-sensitive plate. The variation in heat flux causes an incremental temperature difference to be measured between the

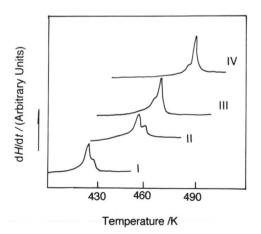

Figure 2.4. DTA heating curves of polyethylene recorded at various pressures. (I) 0.1, (II) 200, (III) 250, and (IV) 350 MPa (courtesy of Y. Maeda)

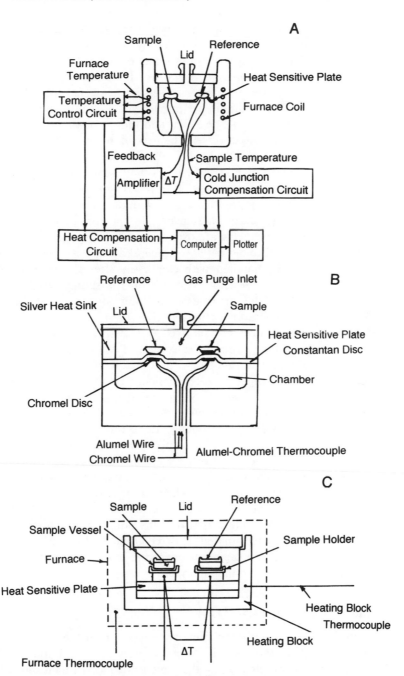

Figure 2.5. (A) Schematic diagram of quantitative DTA apparatus; (B) TA Instruments design (by permission of TA Instruments); (C) Seiko Instruments design (by permission of Seiko Instruments)

heat-sensitive plate and the furnace. The heat capacity of the heat-sensitive plate as a function of temperature is measured by adiabatic calorimetry during the manufacturing process, allowing an estimate of the enthalpy of transition to be made from the incremental temperature fluctuation. The sample and reference material are placed in sample vessels and inserted into the sample and reference holders. For optimum performance of a quantitive DTA system the sample should weigh < 10 mg, be as flat and thin as possible and be evenly placed on the base of the sample vessel. The maximum sensitivity of a quantitative DTA instrument is typically 35 µW.

The furnace of a quantitative DTA system is large and some designs can be operated at temperatures greater than 1000 K. When the sample holder is constructed from platinum or alumina the maximum operating temperature is approximately 1500 K. A linear instrument baseline can be easily obtained because the large furnace heats the atmosphere surrounding the sample holder. The outside of the sample holder assembly casing can become very hot during operation and should be handled with care even after the instrument has finished a scan. Moisture does not condense on the sample holder in sub-ambient mode because of the large heater, making adjustment of the instrument baseline at low temperatures relatively easy.

The time needed to stabilize a quantitive DTA instrument at an isothermal temperature is long for both heating and cooling. For example, when cooling at the maximum programmed rate from 770 to 300 K approximately 30–100 s are required before the temperature equilibrates. To avoid overshooting the temperature on heating the constants of the PID (proportional integral differential) temperature control programme must be adjusted, especially for isothermal experiments where the scanning rate to the isothermal temperature is high. Overshooting also makes heat capacity measurements difficult. The temperature difference between the furnace and the sample can be very large on heating and cooling, particularly if a high scanning rate is used. Instruments are generally constructed so that when the furnace temperature is equal to the selected final temperature the scan is terminated. The sample temperature may be well below this value, and from the point of view of the user the experiment is prematurely terminated. It is recommended to verify the difference between the furnace and sample temperature under the proposed experimental conditions before beginning analysis.

2.3 Triple-cell Quantitative DTA

At high temperatures the effect of radiative energy from the furnace can no longer be neglected as this contribution increases as T^4. Triple-cell DTA systems have been constructed, using the same principle as quantitative DTA, to measure accurately the enthalpy of transition at temperatures greater than 1000 K (Figure 2.6)[1]. A vacant sample vessel, the sample and a reference material are measured simultaneously. The repeatability of the DTA curve

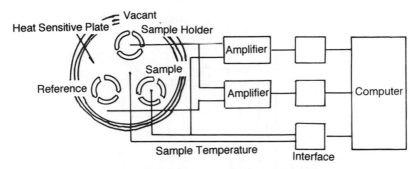

Figure 2.6. Schematic diagram of triple-cell DTA system. High-temperature enthalpy measurements are more precise with this instrument as compared with standard quantitative DTA systems. (Reproduced by permission of Elsevier Science Publishers from *Thermochimica Acta*, **223**, 7 (1993))

with this instrument is ± 3% up to 1500 K. The radiative effect is alleviated by placing three high thermal conductivity adiabatic walls between the furnace and the sample holder assembly.

2.4 Power Compensation Differential Scanning Calorimetry (DSC)

A power compensation-type differential scanning calorimeter employs a different operating principle from the DTA systems presented earlier. The structure of a power compensation type DSC instrument is shown in Figure 2.7. The base of the sample holder assembly is placed in a reservoir of coolant. The sample and reference holders are individually equipped with a resistance sensor, which measures the temperature of the base of the holder, and a resistance heater. If a temperature difference is detected between the sample and reference, due to a phase change in the sample, energy is supplied until the temperature difference is less than a threshold value, typically < 0.01 K. The energy input per unit time is recorded as a function of temperature or time. A simplified consideration of the thermal properties of this configuration shows that the energy input is proportional to the heat capacity of the sample. The maximum sensitivity of this instrument is 35 µW.

The temperature range of a power compensation DSC system is between 110 and 1000 K depending on the model of sample holder assembly chosen. Some units are only designed to operate above 290 K, whereas others can be used over the entire temperature range. Temperature and energy calibration are achieved using the standard reference materials in Appendix 2.1. The heater of a power compensation-type DSC instrument is smaller than that of a quantitative DTA apparatus, so that the temperature response is quicker and higher scanning rates can be used. Instruments display scanning rates from 0.3 to 320 K/min on heating and cooling. The maximum reliable scanning rate is 60 K/min. Isothermal experiments, annealing (single- and multi-step) and heat capacity

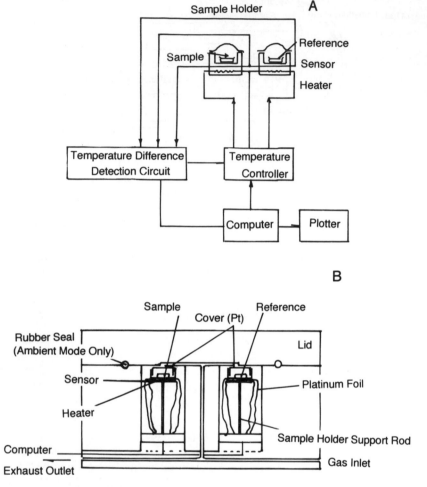

Figure 2.7. (A) Block diagram and (B) schematic diagram of power compensation DSC system (by permission of Perkin-Elmer Corp.)

measurements can be performed more readily using the power compensation-type instrument. Maintaining the instrument baseline linearity is a problem at high temperatures or in the sub-ambient mode. Moisture condensation on the sample holder must be avoided during sub-ambient operation.

2.5 Modulated DSC (MDSC)

A variation on the standard temperature programmes used in TA has recently been introduced. This variation can be applied in principle to quantitative DTA and power compensation DSC systems, and is called modulated (or oscillated)

DSC. A perturbation is applied to the temperature control programme in the form of an oscillating sine wave of known frequency. In the case of quantitative DTA the temperature difference data are collected in the usual manner. A Fourier transformation is performed to extract the data, which are deconvoluted into a heat flow and a temperature signal, both of which contain cyclic components. The phase difference between these cyclic components can be plotted as a function of temperature, in addition to traditional DSC plots. There are a number of problems with this variation of DSC. First, a rigorous theoretical description of the effect of a perturbation in the temperature programme on the dynamics of these systems is not available. Current theoretical models are totally unsatisfactory, assuming among other things that 'the response of the rate of the kinetic process to temperature can be approximated as linear' and that in the presence of modulation the function which describes the kinetics of any transformation in the sample is given by the sum of 'the average underlying kinetic function' plus a sinusoidal contribution which has the same form and frequency as the original modulation of the temperature programme, with no phase difference[2]. From an engineering viewpoint there are also some inconsistencies. MDSC requires no fundamental changes to the instrument. Only software modifications of the temperature control programme and the data processing routines are made. The true modulation frequency and modulation amplitude are limited, within a very narrow range, by the instrument design and not the software. The modulation frequency and amplitude reported by some manufacturers are clearly not possible given their current designs. For the time being, in the absence of a clear methodology for interpretation of results and given the uncertainties over the actual performance of the instrument, results from MDSC should be treated with caution.

2.6 High-sensitivity DSC (HS-DSC)

High-sensitivity DSC systems were originally constructed to measure the denaturation of biopolymers in dilute solution. The transition temperatures of biopolymers range from 265 to 380 K. The enthalpy of denaturation for proteins is typically 6 kJ/(mol amino acid residues). HS-DSC instruments measure approximately 1 ml of sample at a heating rate between 0.1 and 2.5 K/min. Two types of HS-DSC instrument are commercially available. One is based on heat-flux DSC, where the sensitivity is improved by increasing the sample size, using several thermocouples connected in series to measure the sample and reference temperatures, and by increasing the size of the heat sink to minimize temperature fluctuations. The maximum sensitivity of heat-flux HS-DSC systems is between 1.0 and 0.4 μW, depending on the model. The other type is an adiabatic HS-DSC system. A Privalov calorimeter is shown in Figure 2.8 as an example of adiabatic HS-DSC apparatus. Heating elements are placed in the sample and reference holders which are surrounded by two adiabatic shields. The temperatures of the sample and reference are measured

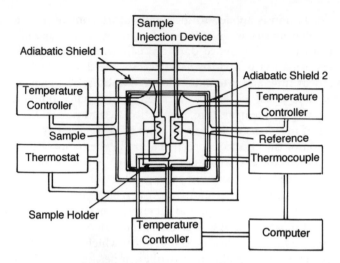

Figure 2.8. Schematic diagram of Privalov adiabatic HS-DSC system (by permission of Russian Academy of Natural Sciences)

and electric current is supplied to the heating elements to minimize any temperature difference. Enthalpy calibration is performed by applying a known amount of electric current and measuring the heat capacity change of a pure water sample or a standard buffer solution. The maximum sensitivity of a Privalov HS-DSC instrument is $0.4\,\mu W$.

2.7 Data Analysis and Computer Software

Commercially available software for TA instruments performs a number of tasks (Table 2.2). Software for more specialized purposes is generally written by users. Data analysis using computer software is more convenient than analysis by hand. However, it is necessary to understand the characteristics of TA data before using the software. When analysing DSC curves software can easily convert a glitch due to electrical noise into a first order phase transition and create a glass transition or a broad peak from the curvature of the sample baseline. Computer software can also generate artifacts in the data through baseline smoothing and baseline correction, in particular. If a large amount of smoothing and/or baseline correction are necessary, it is better to review the sample preparation and the experimental conditions to improve the data rather than correct the data by computer.

2.8 Automated TA Systems

When large numbers of samples are routinely measured, for example in quality control, an automated sample supplier can be fitted to the TA unit. Samples

Table 2.2. Commercially available TA software

TA instrument	Software function
General (DTA, DSC, TG, TMA, DMA)	Variation of signal amplitude
	Signal and temperature calibration
	Accumulation and storage of data
	Baseline smoothing
	Display and calculation of transition temperatures
	Display of multiple curves
	Curve subtraction
	Derivative TA curve
	Baseline correction
DSC	Display and calculation of transition enthalpy
	Heat capacity determination
	Purity calculation
	Reaction rate calculation
TG	Conversion from mass change to % mass change
	Reaction rate calculation
TMA, DMA	Thermal expansion coefficient calculation
	Display of stress–strain curve
	Display of creep curve
	Display of stress-relaxation curve
	Arrhenius plot and associated parameters
	Calculation and display of master curve

are handled and placed in the instrument by a robot arm, and removed after measurements have been completed. At present, robot arms are commercially available for DSC, TG–DTA and TMA instruments. An automated sample supplier is shown in Figure 2.9.

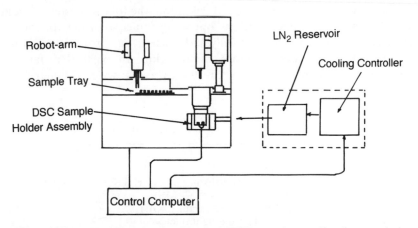

Figure 2.9. Schematic diagram of automated DSC sample supplier (by permission of Seiko Instruments)

2.9 Simultaneous TA Analysis

Various TA apparatus can be combined so that several physical properties can be measured simultaneously. Table 2.3 lists the most commonly available simultaneous analysis instruments.

2.10 Installation and Maintenance

The following points should be considered when installing a DSC (or DTA) instrument in a laboratory. (1) The apparatus should be placed on a level surface approximately 1 m above floor level. (2) The ambient temperature in the vicinity of the instrument should be maintained as constant as possible between 288 and 303 K with a relative humidity < 75%. Where the relative humidity exceeds 75% the instrument should be located in an air-conditioned room. (3) The electric power supply must be stable and a voltage regulator should be used to isolate the apparatus from voltage fluctuations. In the event of a power failure the instrument will shut down, but the switches on the instrument remain in the 'on' position. When the power is restored the instrument modules will power-up in a random sequence and the probability of damaging the instrument is high. The electric supply should be latched so that when the electric supply is restored the latch will ensure that no power is supplied to the instrument until the latch is manually reset by the user at the same time as the other switches on the apparatus are reset. (4) The apparatus should not be in direct sunlight or directly exposed to wind currents (including those from air conditioners). (5) The instrument should be located far from sources of strong magnetic and electric fields, microwaves or other high frequency signals. (6) The instrument should be isolated from mechanical vibrations. In countries where earthquakes occur frequently the apparatus should be placed on a surface which is secured to a load-bearing wall.

To maintain the instrument in good condition, the following steps should be taken. *Beginners:* (1) Understand the operating principle of the apparatus. (2) Read the instruction manuals carefully and discuss your proposed experiments with an experienced user(s) before commencing. In particular, familiarize yourself with those precautions necessary to avoid serious damage to the

Table 2.3.   Simultaneous TA techniques

TG	TG–DTA
	TG–DSC
	TG–fourier transform infrared (FTIR) spectroscopy
	TG–mass spectroscopy (MS)
	TG–Gas chromatography (GC)
DTA/DSC	DTA–polarizing light microscopy
	DTA–X-ray diffractometry

instrument. (4) Immediately contact an experienced user if the apparatus displays any unusual response. *Advanced users:* (1) Record the users name, sample name, date and experimental conditions after each series of measurements has been completed. (2) Maintain a small purge gas flow through the instrument even when it is not in use. (3) Contact your repair engineer immediately in the event of instrument failure.

2.11 References

[1] Takahashi, Y. and Asou, M. *Thermochimica Acta* **223**, 7 (1993).
[2] Reading, M. *Trends in Polymer Science* **1**, 248 (1993).

3 CALIBRATION AND SAMPLE PREPARATION

3.1 Baseline

In DSC (and DTA) a distinction must be made between the recorded baseline in the presence and absence of a sample. The difference is clearly illustrated in Figure 3.1A. By placing empty sample vessels in the sample holder and the reference holder at T_i (320 K) for 1 min the isothermal curve I is recorded. On heating to T_e (400 K) at a constant rate the instrument baseline, curve II, is obtained. Maintaining the empty sample vessels at T_e for 1 min produces curve III. Ideally, curves I, II and III should form a continuous straight line. In practice, curve II deviates from the isothermal curves, the direction and magnitude of the deviation depending on the instrument design and the experimental conditions. If a sample is now placed in the sample vessel and measured under the same experimental conditions, the DSC curve rises linearly owing to the change in heat capacity of the sample as a function of temperature. The linear portion of the DSC curve exhibiting no endothermic or exothermic deviation in the presence of a sample is called the sample baseline. The DSC curves recorded for a 5 mg sample of polystyrene at 5 and 10 K/min are presented in Figure 3.1B. The step-like change in the heat capacity of the sample in the region of 380 K is attributed to the glass transition of polystyrene. That part of the DSC curve outside the transition zone is the sample baseline. Often the sample and instrument baselines are confused since in general only the framed portion of Figure 3.1A is presented.

3.1.1 BASELINE CURVATURE AND NOISE

It is recommended to scan the DSC under the proposed experimental conditions to check the curvature and noise level of the instrument baseline before analysing samples. There are several reasons why the instrument baseline may be curved and/or display a high noise level. Trace amounts of residue from a previous experiment may be attached to the sample holder. Decomposed or sublimated compounds frequently condense on the sample holder, distorting the shape and quality of the instrument baseline. If the instrument baseline is still not satisfactory, following cleaning with ethanol or acetone, there may be a problem with the purge gas flow. The linearity of the instrument baseline will be reduced if the flow rate of the purge gas is not constant or the purge gas

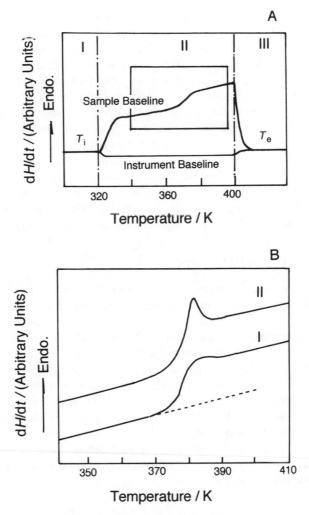

Figure 3.1. (A) Instrument and sample baseline of a power compensation type DSC using polystyrene as a sample. (B) DSC heating curves of polystyrene recorded at (I) 5 and (II) 10 K/min

contains a large amount of water vapour. The mains electric supply is generally not sufficiently stable for a sensitive instrument such as a DSC. Voltage spikes decrease the operating life of the instrument and produce a lot of noise on the instrument baseline. It is recommended to place a voltage regulator between the instrument and the mains supply. The electrical characteristics of the instrument vary slowly with time and readjustment is necessary to maintain a good instrument baseline. An electric malfunction is likely if the instrument baseline is still unsuitable after the instrument electronics have been adjusted over their full operating range. In this case, it is recommended to consult a qualified repair engineer before proceding any further.

The instrument baseline recorded for successive scans, under the same conditions, should be identical. If this is not the case, moisture may have condensed on the sample holder. Increasing the flow rate of dried purge gas should alleviate this problem. In sub-ambient mode, the shape of the instrument baseline is strongly dependent on the level of coolant in the reservoir. This level should be maintained as constant as possible over the entire course of the experiments. If the DSC curve in the isothermal regions before or after the scan is not linear, it is likely that some change in the chemical or physical properties of the sample has occurred at that temperature. For example, polystyrene films containing trace amounts of organic solvent, used in casting the films, exhibit an endothermic deviation in the DSC curve in the isothermal region before the scan due to the vaporization of residual solvent.

3.1.2 BASELINE SUBTRACTION

When estimating a transition temperature and the associated enthalpy change from a DSC curve, it is necessary to extrapolate the sample baseline into the transition region from both the high- and low-temperature sides of the transition. The extrapolation is assumed to be linear (Figure 3.2), except in the case of high-sensitivity DSC instruments (Section 5.13). However, if the instrument baseline has a high degree of curvature, this method is not applicable and the transition, particularly the glass transition, may be difficult to characterize. If a chart recorder has been used to record the DSC curve the instrument baseline should be superimposed on the sample baseline over the entire temperature range of the experiment and the difference area used to characterize the transition. The instrument baseline can be easily subtracted from the DSC curve using the appropriate software option of a computer-controlled instrument.

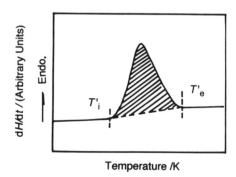

Figure 3.2. Linear extrapolated sample baseline used to calculate the enthalpy of transition

3.1.3 BASELINE CORRECTION

The baseline subtraction option available in DSC software performs a linear subtraction of the instrument baseline from the DSC curve to improve the resolution of the transition of interest. A baseline correction option multiplies the sample baseline by a complex mathematical function in order to improve the linearity of the DSC curve. Baseline correction procedures should be used with extreme caution as the form of the sample baseline contains information on the variation in heat capacity of the sample as a function of temperature. For example, in glassy polymers containing a large proportion of hydrogen bonds the glass transition occurs over a broad temperature interval and is recorded as a gradual endothermic deviation in the DSC curve, in contrast to the sharp step-like glass transition observed with polystyrene (Figure 3.1B). If a baseline correction is applied to the curve of the hydrogen bonded polymer the glass transition disappears. Transition enthalpy and characteristic temperatures determined from a corrected DSC curve are less representative than those of an uncorrected curve.

3.2 Temperature and Enthalpy Calibration

Standard reference materials are used to calibrate the temperature and energy scales of DSC instruments. Following calibration the characteristic temperatures and the enthalpy associated with a phase change can be measured for any sample. The upper limit for the precision of all measurements is set by the accuracy of the calibration. Temperature calibration is generally carried out using the melting temperature of metals whose purity is ⩾99.99% (purity is determined by vapour analysis). The melting point of the high-purity metal is measured by adiabatic calorimetry and a list of suitable metals whose melting temperatures extend over a broad temperature range is presented in Appendix 2.1. Laboratory-grade chemicals are not suitable for use as calibration standards even if the melting temperature is clearly described on the label. The presence of trace amounts of impurities has a large effect on the observed melting temperature (Section 5.6).

At least two standard reference materials, whose transition temperatures span the sample transition interval, should be used to calibrate the instrument. To prepare the reference material a small piece (1–2 mg) of the high-purity metal is cut from the centre of the metal block. The use of metal whose surface has been exposed to air should be avoided as oxidation of the metal surface alters the melting temperature of the substance. Sometimes the standard reference material is supplied as a fine powder which has a large surface area. The temperature and enthalpy of the powder will be different from those of the same purity metal in block form. Temperature calibration should be performed on the heating cycle as significant supercooling of the metal can occur on the cooling cycle rendering temperature calibration difficult.

Furthermore, calibration should be carried out under the same experimental conditions as for the proposed experiment. In particular, the same heating rate should be used as the observed melting temperature is strongly influenced by the temperature gradient between the sample and the sample holder, which depends on, among other things, the heating rate. From the DSC curve the transition temperature and enthalpy of the standard reference material are determined using the procedures detailed in Sections 5.1 and 5.2, respectively. The instrument is then adjusted so that the measured transition temperature and enthalpy of the standard reference material correspond with the accepted values.

Mercury ($T_m = 234.4$ K) and gallium ($T_m = 303.0$ K) are sometimes used as standard reference materials for temperature calibration. Both are poisonous and form alloys when they come in contact with aluminium in the molten state. If molten mercury or gallium is kept in an aluminium sample vessel for a long period it can leak from the sample vessel onto the sample holder and form a metallic alloy, with disasterous consequences for the instrument. When used as calibration standards these metals should be cooled immediately following melting.

Reference material sets which are certified by the International Confederation for Thermal Analysis and Calorimetry (ICTAC) are available through the US National Institute of Standards and Testing (NIST), and are listed in Appendix 2.2. High-purity metals and organic compounds including polymers have been certified. If the standard reference material must be dispensed with a syringe into the sample vessel (for example, cyclohexane), care must be taken to ensure that only one droplet is formed in the sample vessel. Multiple transition peaks will be observed if there is more than one droplet present. The transition temperatures listed in Appendix 2.2 are the statistical mean values of measurements made in a number of laboratories and institutes. The ICTAC reference materials are certified for temperature calibration only and not for enthalpy calibration. The reference temperatures in Appendix 2.1 should be used if very accurate calibration of the instrument is required. In order to determine the heat capacity (C_p) of a sample, sapphire (α-alumina, Al_2O_3) is used as a standard reference material. The C_p of sapphire as a function of temperature is given in Appendix 2.3.

3.3 Sample Vessel

Sample vessels are commercially available in various shapes made from a range of materials, including aluminium, carbon, gold, platinum, silver and stainless steel (Figure 3.3 and Table 3.1). Open-type sample vessels do not seal hermetically even when closed with the sealing press supplied by the manufacturer. Samples which evolve volatile components, sublime or decompose should not be measured using an open-type sample vessel. It is recommended to carry out a preliminary TG analysis of a sample whose decomposition characterisitcs are unknown before commencing DSC measurements.

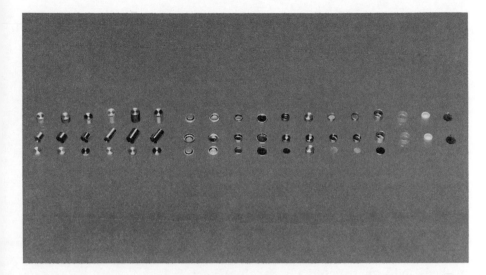

Figure 3.3.   Selection of DSC sample vessels (by permission of Seiko Instruments)

The mass of commercially available sample vessels ranges from 10 to 300 mg. When using an aluminium sample vessel the temperature should not exceed 830 K. If the aluminium sample vessel should melt on the sample holder, alloying will occur and the sample holder will be irreparably damaged. Gold or platinium sample vessels should be used for high-temperature measurements. The maximum safe operating temperatures of various materials used for making sample vessels are listed in Table 3.2. Liquids, gels, bio-materials and other materials likely to produce volatile components should be

Table 3.1.   Selection of sample vessels used in DSC and DTA

Sample vessel			Sample
Type	Material	Shape	
Open	Al, Au, C	Sample	Film, powder block, fibre
Sealed	Al, Ag, Au stainless steel	Rubber ring Sample	Solution Gel Biomaterial Samples which decompose sublime or release volatile solvents

Table 3.2. Operating temperatures of sample vessels made from various materials

Material	Maximum recommended operating temperature/K
Aluminium	830
Aluminium nitride	750 (air)
	1200 (N_2)
Carbon	500 (air)
	1150 (N_2)
	2100 (Ar, He)
Gold	1200
Platinum	1900
Silver	1100
Stainless steel	400

measured using hermetically sealed sample vessels. Several types of sealing methods and sealing presses are available (Figure 3.4).

When analysing a large amount of sample at a slow scanning rate, a silver sample vessel is recommended owing to its high thermal conductivity. A reaction between the solvent and the inner surfaces of a sample vessel can cause unexpected features to be observed on the DSC curve. For example, water in a sample can react with aluminium and an exothermic peak is observed in the region of 400 K. The peak is due to the formation of aluminium hydroxide [$Al(OH)_3$] on the inner surface of the sample vessel. This problem can be circumvented by boiling the sample vessel in water, or sealing it in an autoclave with a small amount of water at 390 K, for several hours. The sample vessel is coated with a layer of aluminium hydroxide and no further reaction takes place during subsequent DSC scans. This procedure does reduce the

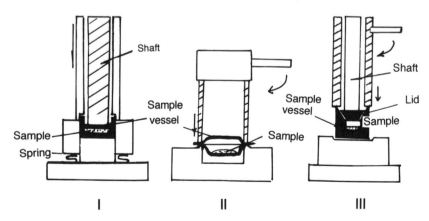

Figure 3.4. Sample vessel sealing presses for use with (I) open-type sample vessels and (II) and (III) hermetically sealed-type sample vessels

ductility of the aluminium and the sample vessel is more difficult to seal here-metically. After the measurements are completed the sample vessel should be reweighed to confirm that no mass loss has taken place.

3.4 Sample Preparation

The sample should in good thermal contact with the base of the sample vessel in order to obtain a reliable DSC curve. This can be achieved by bringing the sample to a temperature higher than the melting or glass transition tempera-ture. The second heating curve, obtained after the sample has been cooled at a programmed rate, has a lower noise level than the first heating curve. If the first heating scan contains important information this procedure is not recommended and the sample should be carefully packed into the sample vessel to ensure good thermal contact. Figure 3.5 presents DSC melting curves of polyethylene. Sample I was crystallized at 470 K under a pressure of 500 MPa. During this high-pressure crystallization extended chain-type crys-tals are formed. The first melting curve records the melting of the extended

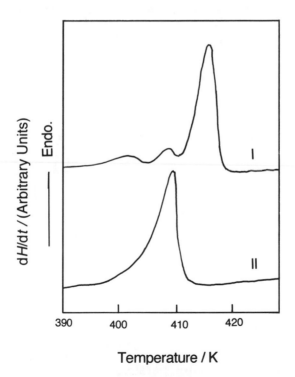

Figure 3.5. DSC melting curves of (I) polyethylene extended chain-type crystals prepared by crystallization at 470 K under 500 MPa pressure and (II) folded chain-type polyethylene crystals prepared under atmospheric pressure

chain-type crystals. Once the extended chain-type crystals have melted it is impossible to re-form the same type of crystal at atmospheric pressure. Folded chain-type crystals are formed during cooling and the second heating curve (II) shows the melting of these crystals.

3.4.1 FILMS, SHEETS AND MEMBRANES

Films, sheets and membranes should be cut into discs according to the diameter of the sample vessel. A sample when cut from a sheet must be chosen carefully in order to be truely representative. Discs should not be cut near the edge or a defect in the sheet. The molecular axis of a polymer is frequently oriented during the stretching and rolling stages of film and sheet production. The first heating curve is markedly affected by the mechanical and thermal history of the sample. The DSC melting curves of a polyethylene film stretched 20 times are presented in Figure 3.6. The stretched film shrinks during heating and a complex melting endotherm is observed. Sometimes it is necessary to measure DSC heating curves of stretched samples while main-taining the oriented molecular structure. Several methods can be employed to prevent film shrinkage during heating. A mechanical device, such as a horseshoe-type spring, can be placed in the sample vessel. Unfortunately, this affects the heat conductivity and increases the thermal mass of the sample vessel. Alternatively, the sample can be chemically cross-linked. The original structure is maintained up to the melting temperature, although the chemi-

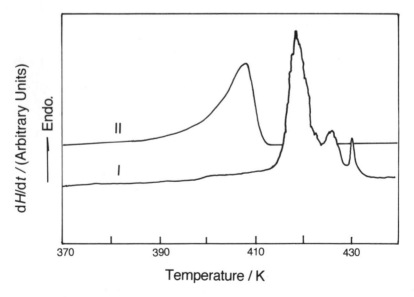

Figure 3.6. DSC melting curves of polyethylene film stretched 20 times. (I) First heating and (II) second heating scan

cally modified structure may not be exactly the same as that of the original sample.

3.4.2 GRANULES AND BLOCKS

Suitable shapes can be cut from granules using a razor blade or microtome. If a large shear stress is applied to the sample during cutting, the high-order structure of the polymer may change, creating a new melting peak. It is recommended to prepare films and sheets from granules or a block. Two methods are generally used to prepare films. The polymer granules can be dissolved in a good solvent at a solution concentration of less than 10 wt-%, and cast on a clean glass plate which is placed on a horizontal surface. The solvent should be evaporated slowly. If the rate of vaporization is high at room temperature, casting should be carried out in a desiccator. Where the rate of vaporization is low, the desiccator is evacuated slowly and the film dried under reduced pressure. The dried film is removed from the glass plate using a razor blade. Films prepared by solvent casting often contain nominal amounts of solvent even after drying. Small amounts of solvent molecules may enhace molecular motion in the polymer, shifting the first-order phase transition and the glass transition to lower temperatures. An endothermic deviation in the sample baseline may be observed due to the vaporization of solvent residue. Even if the solvent is completely removed, the original polymer structure, found in the granule, may be retained and observed in the first heating curve of the film.

Granules, sandwiched between two metal plates, can be pressed under pressure (50–150 kg/cm^2) in the molten state using a hot-stage pressing machine. It is recommended to measure the decomposition temperature of the sample before heat pressing is carried out. Degradation or depolymerization can take place if the sample is maintained for a long time in the molten state. After pressing, the sample can be cooled slowly or quenched to a predetermined temperature. If the films are likely to stick to the metal plates, polytetrafluoroethylene (PTFE) sheets should be used for sandwiching instead of the metal plates. Although reagents are available to remove the film from the plates, it is better not to use them, since they contaminate the sample.

3.4.3 POWDERS

Static electricity may disrupt the packing of powdered samples. When static electricity is a problem, aluminium foil is spread on the work surface and an electric wire connected from the foil to electric ground. Tweezers are available made from PTFE and other non-metallic materials, which also reduces static electicity. Powdered samples should be packed into the sample vessel on the metal foil, ideally in a low-humidity environment.

3.4.4 FIBRES AND FABRICS

Molecular chains are oriented in polymeric fibres. Commercial fibre and fabric samples contain various kinds of contaminants, such as dyes, colour pigments, anti-oxidizing chemicals and light-reflecting additives. Low molecular mass compounds act as a nucleus for crystallization, influencing the crystallization behaviour of fabrics and fibres. When cut into test pieces, the large shear stress caused by cutting can affect the morphology, and therefore the DSC curves, of these materials. The characteristics of the melting transition are effected by stretching which can occur during handling. For example, DSC curves of clothes worn by traffic accident victims made from a poly(ethylene terephthalate)–cotton fabric were found to show a sub-peak beside the main melting peak caused by shear stress. Fibres and fabrics contain a large amount of air whose heat conductivity is low and therefore it is necessary to pack the sample into the sample vessel tightly so that thermal gradients in the sample are minimized.

3.4.5 BIO-MATERIALS AND GELS

Bio-materials and gels contain a large amount of water and often a lot of water is lost during handling. This makes it difficult to evaluate the exact polymer concentration of the sample. If a sample is unaffected by freezing, it can be immersed in liquid nitrogen or another coolant and cut into test pieces while frozen. In order to calculate the exact mass of the sample, the sample vessel should be weighed before and after sealing. Procedures to determine the exact water content of a sample are described in Section 5.11.1.

3.4.6 STORING SAMPLES

The conditions under which a sample is stored influence its high-order structure. Enthalpy relaxation proceeds at temperatures lower than the glass transition temperature of amorphous or semi-crystalline polymers. Such samples may become hard and brittle with time. The relaxation can be observed by DSC as the increase in the endothermic peak at the glass transition (Section 5.4.1). When kept in moist conditions, some polymers crystallize in the presence of absorbed solvent (solvent-induced crystallization). Amorphous poly(ethylene terephthalate) easily crystallizes in the vapour of organic solvents such as benzene and toluene, and should be kept in a desiccator with drying chemicals (silica gel or diphosphorus pentaoxide). If necessary, the desiccator can be kept in a refrigerator. Many polymers degrade when irradiated by ultraviolet rays, and photosensitive polymers should be stored in the dark. Natural polymers are damaged by biodegradation and should be stored under sterile conditions. Under moist, warm conditions, micro-organisms may breed in natural polymer samples. Atmospheric oxygen also affects the stability of polymers. Where a sample is to be storage for a long time the storing conditions must be carefully chosen according to the nature of the sample.

3.5 Temperature Gradient in Sample

During heating at a programmed rate, temperature gradients are developed between the furnace and the sample, and within the sample itself. The magnitude of the temperature gradients increases with increasing mass of the sample and the scanning rate. The melting temperature of a high-purity indium sample sandwiched between polyethylene films has been measured as a function of the distance from the bottom of the sample vessel to the indium and as a function of heating rate. The results are summarized in Figure 3.7. It can be seen that the difference between the observed and true melting temperatures of the indium sample increases with increasing heating rate and thickness of the polyethylene layer. A temperature gradient also exists between the temperature sensor of the sample holder and the bottom surface of the sample vessel. The characteristics of this temperature gradient depend on the design of the sample holder, the shape and material of the sample vessel and the type of DSC instrument.

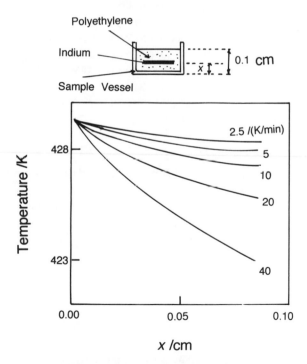

Figure 3.7. Melting temperature of In, sandwiched between polyethylene films, as a function of the distance from the base of the sample vessel to the In sample (x) and as a function of heating rate (courtesy of S. Ichihara)

3.5.1 MASS OF SAMPLE

It is easier to obtain data free from the effects of temperature gradients with low-mass samples, where the true sample temperature can be more precisely defined. Generally when measuring a first-order phase transition at a scanning rate of 5–10 K/min, 1–5 mg of sample is used. The melting temperatures of two samples of a polyethylene film weighing 0.25 and 2.5 mg as a function of heating rate are presented in Figure 3.8. The peak melting temperature of the low-mass sample is approximately constant with heating rate, whereas an increase in melting temperature is observed for the heavier sample. The DSC heating curves of *n*-hexacosane (an *n*-alkane) measured at 10 K/min exhibit two endotherms (Figure 3.9). The low-temperature peak is associated with a crystal-to-crystal transition and the high-temperature peak with melting. Plotting the observed transition temperatures as a function of sample mass, over the range 0.5–100 µg reveals that the transition temperatures are constant in this low sample mass range. Commercially available DSC instruments are sufficiently sensitive to record phase changes in samples in the region of 0.001 mg.

The sample must be representative of the material as a whole, and the lower limit of the sample mass may be determined by the homogeneity of the sample. This is especially true of composites and polymer blends, where frequently 10 mg of sample are required before a representative sample can be prepared. When a large amount of sample is to be analysed the heating rate should be decreased accordingly to compensate for the increased temperature gradients in the sample.

A precise knowledge of the sample mass is not necessary to estimate the transition temperature. In order to measure the transition enthalpy the sample mass should be known to within ± 1 or ± 0.1 µg, if possible. During the

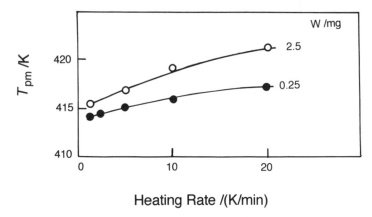

Figure 3.8. Melting temperature of polyethylene film as a function of heating rate. Sample mass; (●) 0.25 and (○) 2.5 mg

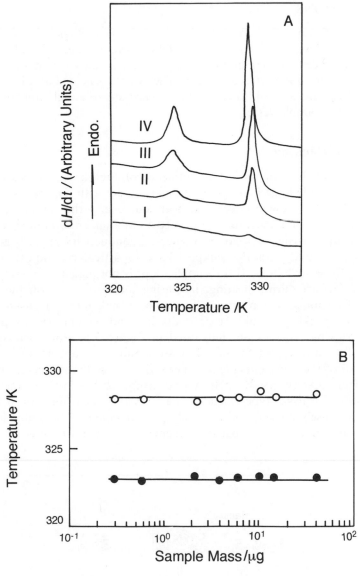

Figure 3.9. (A) DSC heating curves of *n*-hexacosane, recorded at 10 K/min, as a function of sample mass: (I) 0.5; (II) 2.4; (III) 4.1; (IV) 6.2 μg. (B) Transition temperature as a function of sample mass: (●) crystal to crystal transition and (○) melting

weighing procedure the moisture absorbed by the sample is clearly registered at this level of microbalance sensitivity. Hydrophilic polymers can absorb 5–15 wt-% of water, relative to the dry wt of the sample, and this sorbed water can have a dramatic effect on the transition behaviour of the sample.

3.5.2 SOLUTIONS

Low concentration solutions (<0.5 wt-%) of biological samples are commonly analysed by high-sensitivity DSC. Several hundred milligrams of sample are placed in the sample vessel and an equal amount of the pure solvent is placed in the reference vessel. The sample is heated at a low heating rate (<2.5 K/min) to avoid thermal interference due to circulating convection currents in the sample vessel.

3.6 Sample Packing

Sample packing has an important influence of the characteristics of DSC curves. The sample must be in good thermal contact with the sample vessel to optimize the heat flow between the heat source and the sample, reducing thermal lag. The sample should be packed to minimize the number of voids between sample particles, because the thermal conductivity of air is generally very low compared with the sample. This is particularly important when measuring fibres or fabrics. If the sample shape is not regular, the sample may become deformed during heating, increasing the noise level on the sample baseline. This noise is often large enough to be mistaken for a transition peak. The sample packing should be reproducible and as uniform as possible. Various packing schemes are illustrated in Figure 3.10, for open-type and hermetically sealed-type DSC sample vessels. Samples I and IV are correctly packed. All of the other packing schemes will give rise to artifacts in the DSC curve. Figure 3.11 shows the melting peak temperature of polyoxymethylene, in the form of whisker-type extended-chain crystals, as a function of heating rate, for a densely packed and a loosely packed sample. The temperature of melting is dependent on the packing scheme over the entire range of heating rates.

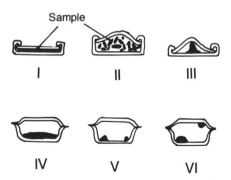

Figure 3.10. Packing schemes for open-type and hermetically sealed-type sample vessels. Only samples I and IV are correctly packed

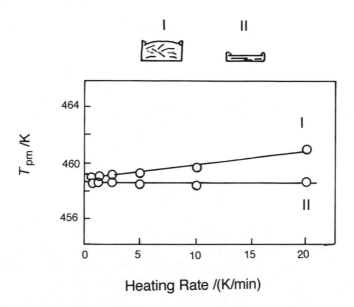

Figure 3.11. Melting temperature of polyoxymethylene as a function of heating rate for (I) loosely packed and (II) densely packed samples

3.6.1 HYDROPHILIC SAMPLE

Hydrophilic samples adsorb moisture from the air and increase in mass during handling. A sample stored in a desiccator with drying agents immediately increases in mass when exposed to humid air. The mass increase is most pronounced in the first 1–5 min, before levelling off gradually. Handling, including packing, should be performed in a dry-box under dry N_2, if a dry sample of a hydrophilic ploymer is required. Alternatively, a large polyethylene bag filled with a dry inert gas (Ar, N_2) will suffice as a preparation chamber.

3.6.2 LIQUID SAMPLE

A single drop of a liquid sample dispensed using a microsyringe should be placed in the centre of the sample vessel. When the sample is separated into two or more droplets, as shown in Figure 3.10 (IV and V), multiple transition peaks may be observed at different temperatures. After sealing, a sample vessel containing a liquid should be kept horizontal during subsequent handling. It is recommended to confirm that no mass loss has occurred during scanning by reweighing the sample vessel after the measurement. An endothermic deviation in the DSC curve, due to solvent vaporization, is observed if the sample vessel is not properly sealed. If the gradient of DSC curve is far greater than that of the instrument baseline, the sample vessel may not be sealed. In this case, the experiment should be immediately terminated to avoid damaging the sample holder.

3.7 Purge Gas

When operating above ambient temperature, N_2 gas is the most commonly employed purge gas. A number of grades of N_2 are available. Standard N_2 gas has a purity of 99.99%, where water vapour and oxygen are the principal impurities. This grade of N_2 should be dried, by passing it through a drying column containing silica gel, before the gas purges the instrument sample holder assembly. Three other grades denoted, A, B and S, are available with purities of 99.9998, 99.9995 and 99.9999%, respectively. Air is not used as a purge gas because of its complex composition and high water vapour content. The nature of the purge gas has an important influence on the shape of the DSC peak, particularly for the decomposition reactions observed in simultaneous TG–DTA.

A number of inert purge gases are used when operating at sub-ambient temperatures. The choice of purge gas is largely determined by the proposed experimental parameters. Table 4.1 lists the thermal conductivities of the most commonly used sub-ambient purge gases (Ar, He and N_2) as a function of temperature and pressure. If the purge gas is changed the temperature and energy scales of the instrument must be recalibrated. In some apparatus the sample holder assembly is housed in a dry box through which N_2 is purged.

The gas flow rate influences the characteristics of the instrument baseline. When the flow rate is too high the instrument baseline becomes unstable. Fluctuations in the flow rate alter the baseline gradient. A purge gas flow rate in the range 20–50 ml/min is recommended. Measurements under reduced pressure are generally only performed with simultaneous TG–DTA instruments and are discussed in Section 4.5.

3.8 Scanning Rate

The height of the sample baseline increases with increasing scanning rate owing to the temperature gradient in the sample. In addition, the melting peak area increases with increasing heating rate. These features are illustrated in Figure 3.12A, which shows the DSC melting curves of poly(ethylene terephthalate) recorded at various heating rates. If the enthalpy of melting of a sample varies with heating rate, all other conditions remaining constant, it is likely that either a chemical reaction takes place during scanning or the high-order structure of the sample changes in the course of the experiment. The kinetics of these processes are dependent on the heating rate. The high-temperature melting peak in Figure 3.12A, which evolves with increased heating rate, is a result of molecular reordering which causes crystallization to occur with heating. The effect of thermal history on the molecular structure of polymers is discussed in Chapter 5.

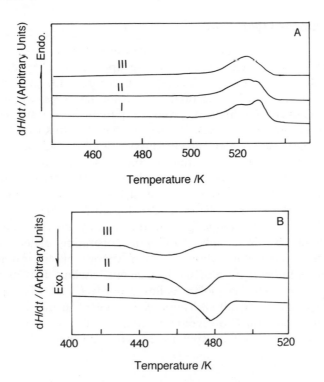

Figure 3.12. (A) DSC melting curves of poly(ethylene terephthalate) as a function of heating rate: (I) 10; (II) 20; (III) 40 K/min. (B) Corresponding cooling curves recorded at (I) 10, (II) 20 and (III) 40 K/min

The shape and characteristic temperatures of a melting peak depend on the heating rate. It is important to state the heating rate when reporting results. The heating rate can be varied from 0.1 to 40 K/min using standard apparatus. However, heating rates in excess of 40 K/min for quantitative DTA instruments and 60 K/min for power compensation-type DSC instruments cannot be accurately controlled irrespective of the heating rate indicated by the instrument. The instrument displays the programmed heating rate and not the true heating rate of the sample. The DSC cooling curves of the same poly(ethylene terephthalate) film cooled at various cooling rates are presented in Figure 3.12B. The temperature of crystallization increases and the width of the crystallization peak decreases with decreasing cooling rate. The problem of reliable fast cooling is even more acute than that of fast heating. The maximum controlled cooling rate is 40 K/min with commercial instruments. However, with some quantitative DTA instruments cooling in excess of 10 K/min is not accurately controlled owing to the large furnace size. Again, instruments only display the programmed cooling rate and not the true cooling rate.

3.9 Sub-ambient Operation

When measurements are made at temperatures below 310 K the sample holder assembly is cooled using a cooling apparatus. The most common types of cooling apparatus are illustrated in Figure 3.13. In Figure 3.13 (I and II) the coolant can be one of the following mixtures, where the lower operational temperature limit is given in parentheses: water–crushed ice (285 K), dry-ice–acetone (240 K), dry-ice–methanol (240 K) or liquid nitrogen (150 K). Dry-ice is solid carbon dioxide and should be crushed into small pieces before mixing to minimize the noise level on the instrument baseline. When using a dry-ice–acetone mixture, the dry-ice must not be allowed to evaporate completely because of the toxicity of acetone vapour. In the pressurized Dewar configuration in Figure 3.13 (III), liquid nitrogen is exclusively used. The purge gas in the vicinity of the sample holder assembly should be thoroughly dried when operating under sub-ambient conditions to avoid condensation of water vapour on the sample holder, which increases the curvature of the instrument baseline. It is recommended to use dry helium when the coolant is liquid nitrogen and a fast cooling rate to low temperatures is required. The level of coolant in the reservoir should be maintained constant to avoid drifting of the instrument baseline. Computer-controlled cooling systems are recommended for experiments

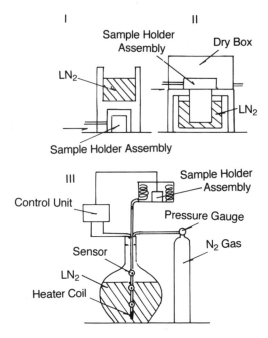

Figure 3.13. Types of DSC cooling apparatus. The appropriate coolants are described in the text

which require very flat instrument baselines (iosthermal crystallization and heat capacity measurements). These units tend to consume a lot of coolant. Temperature and enthalpy calibration of the instrument must be performed each time the coolant is changed. It is recommended that the operating temperature should not exceed 390 K when liquid nitrogen coolant is being used.

4 THERMOGRAVIMETRY

4.1 Introduction

Thermogravimetry (TG) is the branch of thermal analysis which examines the mass change of a sample as a function of temperature in the scanning mode or as a function of time in the isothermal mode. Not all thermal events bring about a change in the mass of the sample (for example, melting, crystallization or glass transition), but there are some very important exceptions which include desorption, absorption, sublimation, vaporization, oxidation, reduction and decomposition. TG is used to characterize the decomposition and thermal stability of materials under a variety of conditions, and to examine the kinetics of the physico-chemical processes occurring in the sample. The mass change characteristics of a material are strongly dependent on the experimental conditions employed. Factors such as sample mass, volume and physical form, the shape and nature of the sample holder, the nature and pressure of the atmosphere in the sample chamber and the scanning rate all have important influences on the characteristics of the recorded TG curve.

TG cannot be considered as a black box technique where fingerprint curves are obtained irrespective of the experimental conditions. Establishing the optimum conditions for TG analysis frequently requires many preliminary tests. It is essential for accurate TG work that the experimental conditions be recorded and that within a given series of samples the optimum conditions be standardized and maintained throughout the course of the experiments. Only then can TG curves from different experiments be compared in a meaningful way.

TG curves are normally plotted with the mass change (Δm) expressed as a percentage on the vertical axis and temperature (T) or time (t) on the horizontal axis. A schematic representation of a one-stage reaction process observed in the scanning mode is shown in Figure 4.1. The reaction is characterized by two temperatures, T_i and T_f, which are called the procedural decomposition temperature and the final temperature, respectively. T_i merely represents the lowest temperature at which the onset of a mass change can be detected for a given set of experimental conditions. Similarly, T_f represents the lowest temperature by which the process responsible for the mass change has been completed. The values of T_i and T_f have no absolute significance as both the reaction temperature and the reaction interval ($T_i - T_f$) have no definitive value but depend on the experimental conditions.

Interpretation of TG data is often facilitated by comparison with data from

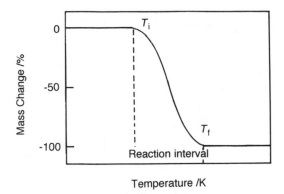

Figure 4.1. Schematic single stage TG curve

other experimental techniques. Many TA instrument manufacturers offer simultaneous TG–DTA apparatus. The advantage of simultaneous apparatus is that the sample and experimental conditions are identical, and therefore directly comparative data can be quickly obtained. However, the performance of one or both components may be compromised in a simultaneous apparatus owing to the contrasting design features of the instruments.

The gaseous products evolved during a TG measurement are a rich source of information and these gases are readily analysed by coupling an appropriate instrument to the TG apparatus. This form of thermal analysis is often referred to as evolved gas analysis (EGA), and is discussed in Section 6.1. Mass spectrometers (TG–MS), Fourier transform infrared spectrometers (TG–FTIR) and gas chromatographs (TG–GC) may be coupled for simultaneous TG–EGA.

4.2 Thermobalance

TG curves are recorded using a thermobalance. The principle elements of a thermobalance are an electronic microbalance, a furnace, a temperature programmer and an instrument for simultaneously recording the outputs from these devices. A thermobalance is illustrated schematically in Figure 4.2.

4.2.1 INSTALLATION AND MAINTENANCE

The location of a thermobalance in the laboratory is of particular importance. The instrument must be isolated from mechanical vibrations. This can be achieved by placing the thermobalance on an isolation table. However, isolation tables are expensive and a more economical solution is to stack polystyrene foam blocks approximately 1 m high with a steel plate (2–3 cm thick) across the top, as an isolation stage. This will isolate the thermobalance from

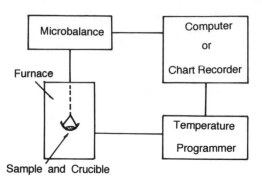

Figure 4.2.   Block diagram of a thermobalance

vibrations in all but the most extreme cases. Alternatively the thermobalance can be placed on a stone slab, preferably with rubber blocks (2–3 cm thick) between the slab and the legs of the thermobalance. The thermobalance should not be exposed to strong air currents, high levels of humidity or large fluctuations from ambient temperature (> 10 K). A sheltered location on an isolation table in a climate-controlled room is ideal. Failure to locate the thermobalance correctly can result in erroneous data and/or loss of resolution due to increased noise. Calibration checks will become more frequent and less reliable under adverse conditions.

The thermobalance must be horizontal on whatever surface it is placed. A spirit-level is normally used for this purpose. Note that it is in the region of the balance mechanism that the spirit-level should be placed. TG instruments are particularly sensitive to improper handling and the manufacturers' instructions should always be consulted before moving the thermobalance. When running an unknown sample or a sample whose decomposition products are unknown, it is recommended to remove all evolved gases safely from the laboratory for proper disposal until such time as the gases can be identified by some form of evolved gas analysis. If the thermobalance has been idle for two or more days the instrument should be given at least 1 h, at operational gas flow rates, to equilibrate before use. Finally, it is useful to keep a log-book containing calibration settings, atmospheric conditions, samples studied and the temperature range of each experiment. In the event of instrument breakdown such a log-book is frequently invaluable to the repair engineer.

4.2.2 MICROBALANCE AND CRUCIBLE

Various designs of microbalance are commercially available including beam, spring, cantilever and torsion balances (Figure 4.3). The microbalance should accurately and reproducibly record the change in mass of the sample, under a range of atmospheric conditions and over a broad temperature range. The

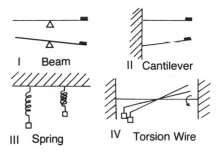

Figure 4.3. Various types of microbalance. (I) beam, (II) cantilever, (III) spring; (IV) torsion wire

microbalance must also provide an electronic signal which can be used to monitor the mass change using a chart recorder or a microcomputer. The sensitivity and range of the microbalance depend on the model chosen, but typically the sensitivity is $\pm 1\,\mu g$ with a maximum sample mass of 100 mg.

It is necessary to calibrate the microbalance periodically. The method of calibration can be either mechanical or electronic, depending on the instrument. Mechanical calibration involves the use of standard precision weights (from 10 to 100 mg) which are placed on the microbalance and the electronics of the microbalance readjusted so that the mass recorded by the microbalance corresponds exactly with the standard mass over the entire mass range of the microbalance. The second method of calibration uses an internal reference weight as the weighing standard and the microbalance can be calibrated at the touch of a button or by selecting the appropriate software option. Mass calibration should be carried out in the absence of gas flow in order to prevent disruption of the calibration by bouyancy and/or convection effects. The thermobalance is normally equipped with a clamp to hold the microbalance arm firmly during loading. After releasing the clamp, the zero point of the balance must be reset. Many users prefer to use the clamp for transporting the thermobalance only and not for routine loading because of the effect on the zero point. In this case clamping is performed manually using tweezers with little or no effect on the zero point. Once a sample has been loaded the microbalance will require 15–30 min to settle down before any measurements can be made.

The sample to be studied is placed in the sample holder or crucible, which is mounted on (or suspended from) the weighing arm of the microbalance. A variety of crucible sizes, shapes and materials are used (Figure 4.4). The melting point of the crucible should be at least 100 K greater than the temperature range of the experiment and there must be no chemical reaction between the crucible and the sample. Crucibles are typically made from platinum, aluminium, quartz or alumina (a ceramic), but crucibles made from other materials are available. The crucible should transfer heat as uniformly and as efficiently as possible to

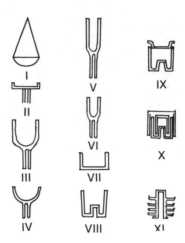

Figure 4.4.   Selection of crucibles used in TG

the sample. The shape, thermal conductivity and thermal mass of the crucible are therefore important. Flat crucibles with a small lip are recommended for powdered samples, but if swelling or spattering is likely then a walled crucible is better. Liquid samples also require a walled crucible. The form of the crucible will to a large extent determine the temperature gradients in the sample. The TG curves recorded for a powdered poly(methyl methacrylate) (PMMA) sample placed in an aluminium crucible and a second identical sample placed in a crucible whose thermal mass was increased by inserting an aluminium disc in the bottom of the crucible, are shown in Figure 4.5A. The shapes and residue rates of the curves are different. The temperature at which the rate of mass loss is a maximum is increased by 5 K using the crucible of higher thermal mass. Simultaneous DTA measurements, presented in Figure 4.5B, reveal a single endotherm in the region of 590 K for the sample placed in the normal crucible, whereas two endotherms are observed in the same temperature region with the crucible of greater thermal mass.

The crucible should be thoroughly cleaned before use by rinsing in methanol and subsequently heating to its maximum operating temperature, in a very high flow rate (150 ml/min) air atmosphere, to burn off volatile contaminants. Periodic cleaning of the sample chamber housing using the same procedure is recommended. The counter balance of the microbalance should be adjusted according to the mass of the crucible used. Typically this involves the addition (or removal) of small weights from the counter balance or adjustment of the position of the counter weight on the balance arm. Compensation for changes in crucible mass are important if a small mass change is expected. Crucibles of equal size decrease in mass in the order Pt > Al > alumina. The use of a crucible cover is not recommended unless decomposition under specialized atmospheric conditions is of interest (Section 4.5).

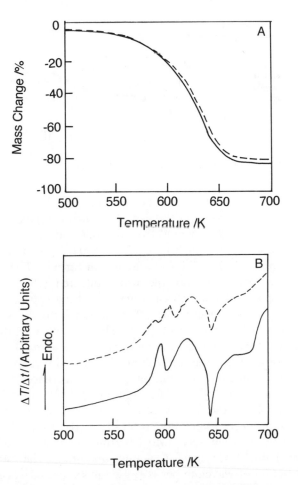

Figure 4.5. (A) Effect of crucible thermal mass on the TG curves of powdered PMMA. Solid line, aluminium crucible (type VII in Figure 4.4); dashed line, aluminium crucible with aluminium disc insert. (B) DTA curves recorded simultaneously. Experimental conditions: 5 K/min heating rate; 5 mg sample; dry N_2 purge gas flow rate, 20 mL/min

4.2.3 FURNACE AND TEMPERATURE PROGRAMMER

The furnace should have a hot zone of uniform temperature which is large enough to accommodate the sample and crucible. The temperature of the hot zone generally corresponds to the recorded furnace temperature. Within the furnace it is inevitable that temperature gradients exist in the vertical and radial directions. These temperature gradients can have a strong influence on the atmosphere in the furnace chamber. Many configurations of microbalance and furnace are available, and some of these are illustrated in Figure 4.6. The

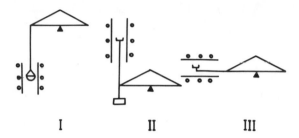

Figure 4.6.   Configurations of microbalance and furnace

furnace heating coil should be wound non-inductively to avoid magnetic interactions between the coil and the sample, which can give rise to spurious apparent mass changes. Coils wound from various materials are commercially available and these include nichrome, $T < 1300\,K$; platinum, $T > 1300\,K$; platinum–10% rhodium, $T < 1800\,K$; and silicon carbide, $T < 1800\,K$.

Thermobalances are also available using infrared (IR) furnaces. These furnaces are composed of symmetrically arranged tungsten filament wires in a reflective housing which focuses the IR radiation onto the sample. An IR furnace can routinely be used up to 1800 K. The principle operating advantage of these furnaces occurs at high temperatures in vacuum where the heating effect is mainly radiative. Under these conditions a coil furnace acts as a black body emitter. The intensity distribution of an IR furnace as a function of applied voltage and the intensity distribution of a black body emitter are shown in Figure 4.7A and B, respectively. It can be seen that for high-temperature operation (*ca* 1300 K) the intensity distribution of the IR furnace is shifted to shorter wavenumbers. Most thermobalances are equipped with a quartz sample chamber housing for vacuum operation. The transmittance of quartz as a function of wavenumber is shown in Figure 4.8. For radiation of wavenumber between 0.2 and 4.0 μm, quartz is essentially transparent. Outside this range the transmittance is approximately zero. If this is compared with the intensity distributions of the coil and IR furnaces, it is clear that while a large fraction of the output from the coil furnace is absorbed, almost the entire output from the IR furnace is transmitted by the quartz. Therefore, under conditions of high temperature and vacuum, in the presence of a quartz housing, temperature control is more reliable using an IR furnace. IR furnaces also have the advantage of very fast true heating rates (up to 1000 K/min). The nominal lifetime of a tungsten filament is 3000 h for $T < 1600\,K$, and less for higher temperatures.

To ensure that the furnace temperature is accurately controlled throughout the experiment, the maximum operating temperature of the furnace should be at least 100 K greater than the desired experimental range. The furnace heat capacity should be as low as possible to facilitate rapid heating of the sample and to reduce thermal lag between the programmed temperature and the true

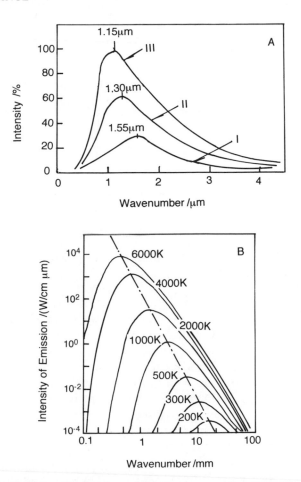

Figure 4.7. (A) Intensity distribution of IR furnace as a function of wavenumber at various heater voltages: (I) 50; (II) 75; (III) 100% of maximum heating voltage. (B) Intensity distribution of a black body emitter as a function of wavenumber at various emitter temperatures

sample temperature. It is essential that the furnace in no way affects the microbalance mechanism. At high temperatures radiation and convection effects can disrupt the microbalance. Various geometries and configurations of microbalance, coupled with the use of radiation shields and convection baffles, are employed to combat these effects.

The operation of the furnace is controlled by a temperature programmer which ensures a linear heating profile over a range of heating rates. A variety of electronic control circuits based on the PID (proportional integral differential) principle are used to provide linear heating profiles. The parameters of the PID circuit are normally set by the manufacturer for optimum perfor-

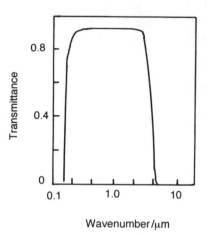

Figure 4.8. Transmittance of quartz as a function of wavenumber

mance under the standard operating conditions of a dry nitrogen atmosphere at 10^5 Pa and a heating rate of 10 K/min. However, when working well outside these conditions, particularly at high heating rates or under isothermal conditions, it is advisable to check the temperature profile of the furnace for overshooting which can significantly raise the sample temperature above the programmed value, sometimes with disastrous results. Many instrument software packages allow the PID parameters to be altered to avoid this problem.

Even with the best furnace and temperature programmer design there is a difference between the the furnace temperature (T_r) and the sample temperature (T_s). This temperature difference is influenced by a number of factors. A thermocouple placed anywhere within the hot zone can record T_r, but measurement of T_s is more difficult. Owing to the nature of the microbalance the thermocouple which monitors T_s is often not in direct contact with the sample, but instead measures the temperature of the atmosphere in close proximity to the sample. Even in the absence of thermal gradients within the sample the true sample temperature is unknown. Fluctuations in sample temperature due to heat evolution in the sample can remain undetected under these conditions. Figure 4.9 illustrates some of the configurations of sample and thermocouple commonly employed.

The nature of the atmosphere in the sample chamber can affect the measured value of T_s because not all gases have the same thermal conductivity at a given temperature and pressure (Table 4.1). At atmospheric pressure in air the measured value of T_s recorded using a thermocouple 5 mm from the sample is *ca* 10 K greater than that measured at the sample. Under the same conditions in a helium atmosphere the difference is *ca* 1 K. Crucibles of large thermal mass or inappropriate geometry can also influence the difference between T_s and T_r, especially at high heating rates. These difficulties can be

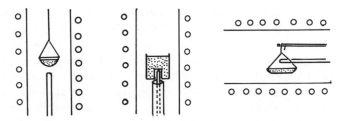

Figure 4.9. Schemes for positioning the sample thermocouple

alleviated by insertion into the crucible of a thermocouple to measure T_s. This is possible by making use of the tolerance in a fine coil wound at the bottom of the thermocouple. Such an arrangement does reduce the microbalance sensitivity to $\pm 2\,\mu g$, but other than this has no detrimental effect on the working of the microbalance.

4.2.4 DATA RECORDING UNIT

The outputs from the microbalance, the furnace and the temperature programmer are recorded using either a chart recorder or a microcomputer (sometimes referred to as a work station). Depending on the number of input channels available the chart recorder can monitor T_r, T_s, Δm and $\Delta m/\Delta t$. In the case of simultaneous TG–DTA another input channel is required to record the DTA signal. The shape of a TG curve is affected by the chart recorder speed. When the chart speed is too slow, mass loss steps appear steep with inflection points and plateaux that are difficult to resolve, especially for overlapping reactions. On the other hand, measurements made from TG curves recorded at high chart speeds tend to give low values for the rate of mass change. A good rule of thumb is 25 cm/h for a heating rate of 5 K/min.

Table 4.1. Thermal conductivities of gases commonly used in TG as a function of temperature and pressure

Pressure/MPa	T/K	Thermal conductivity/mW/m K			
		Air	Ar	He	N_2
0.1	300	26.23	17.77	152.7	25.91
	400	33.28	22.44	188.2	32.49
	500	39.71	26.63	221.2	38.64
1.0	300	26.59	18.09	153.0	26.37
	400	33.54	22.67	188.5	32.83
	500	39.92	26.81	221.4	38.92
10	300	31.21	22.06	156.8	31.48
	400	36.67	25.30	191.2	36.40
	500	42.31	28.79	223.6	41.70

The clear advantage of a microcomputer over a chart recorder is that the former comes equipped with custom software which allows data to be saved, rescaled and replotted, and performs mathematical operations. Multiple TG curves may be plotted simultaneously in addition to comparative plots from other thermal analysis instruments. The most serious drawback to the use of microcomputers is the limited memory size allotted to record each curve. As a result, a TG curve is composed of a finite number of data points irrespective of the duration of the experiment. For short runs there is no resolution problem, but for experiments over several hundred degrees at a slow heating rate the computer may only register a data point once every several seconds. In this case the plotted TG curve is seen to be composed of short, straight segments where only the starting point of each segment is a true data point. Resolution can be seriously reduced under these conditions. Typical *recording* intervals range from a minimum of 0.1 to 1000 s depending on the software and the experimental conditions. Users are recommended to check how many data points per degree will be recorded for a given set of experimental parameters and if necessary these parameters should be altered to optimize the recording conditions either by concentrating on a shorter temperature interval, increasing the heating rate or preheating the sample to a temperature just below the initial decomposition temperature. Chart recorders do not suffer from this drawback as they plot each data point as it is *sampled* by the instrument, which is normally every 0.1 s irrespective of how long the experiment will last.

4.3 Temperature Calibration

The transition used to calibrate the temperature scale of a thermobalance should have the following properties[1]: (i) the width of the transition should be as narrow as possible and have a small energy of transformation; (ii) the transition should be reversible so that the same reference sample can be used several times to check and optimize the calibration; (iii) the temperature of the transition should be independent of the atmospheric composition and pressure, and unaffected by the presence of other standard materials so that a multi-point calibration can be achieved in a single run; and (iv) the transition should be readily observable using standard reference materials in the milligram mass range. Transitions or decompositions which involve the loss of volatile products are usually irreversible and controlled by kinetic factors, and are unsuitable for temperature calibration. Dehydration reactions are also unsuitable because the transition width is strongly influenced by the atmospheric conditions.

One solution is the use of Curie points in ferromagnetic materials. The Curie point of a ferromagnetic material can be broadly described as the temperature at which the ferromagnetic material becomes paramagnetic and the measured magnetic force is effectively reduced to zero. If a ferromagnetic material is placed in a thermobalance crucible, which itself is in a constant magnetic field, the microbalance will register a mass which is the sum of the true mass of the material and a magnetic mass due to the interaction between the ferromagnet and the permanent magnet field. As the ferromagnet is heated through its Curie point the magnetic mass decreases to zero and the balance registers an apparent mass loss (or mass gain, depending on the configuration). If several ferromagnetic standard reference samples are placed in the crucible at once a multi-point calibration curve can be obtained over a wide temperature range (Figure 4.10). Temperature calibration using ferromagnetic alloys is accurate to ±4 K. Magnetic transition can be supercooled and if required calibration in the cooling mode should only be carried out at very slow scanning rates. During the first heating cycle mechanical stress is often released, making the result of the first calibration run unreliable. The standard reference material should be preheated prior to calibration. The biggest disadvantage of Curie point temperature calibration is that the Curie point is sensitive to the composition and purity of the ferromagnetic material. Calibration should only be performed using standard reference materials (Appendix 2.2). Owing to the low mass of the standard reference material, a relatively strong permanent magnetic field may be required before the change in magnetic mass is observed.

A second method of temperature calibration[2] involves suspension of a platinum wire over a boat-shaped crucible by spot welding (Figure 4.11A). To this wire a hook, made of a material of well characterized melting point, is attached which supports a platinum weight. When the melting point of the hook is reached the weight falls, causing an action–reaction response in the microbalance which produces a blip in the baseline and not a mass loss. The blip in the baseline is used as the calibration point (Figure 4.11B). In a variation of this method, a hole is cut in the crucible so that when the weight drops

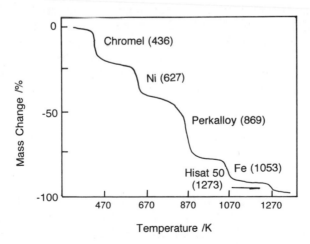

Figure 4.10. Multi-point temperature calibration curve using the Curie point method

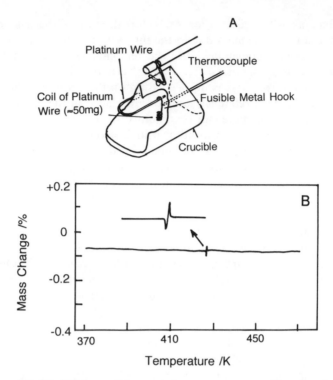

Figure 4.11. (A) Modified crucible used for temperature calibration by the drop method. (B) TG curve exhibiting the calibration 'blip'. (Reproduced by permission of Elsevier Science Publishers from *Thermochimica Acta*, **67**, 241 (1983))

a mass change is recorded. Temperature calibration of $\pm 2\,K$ up to $1500\,K$ is reported using these methods. The difficulties in using drop methods include the following: (i) the hook material must have a low surface tension in the liquid phase; (ii) some soft metals stretch before melting so that the platinum weight may come to rest on the crucible before melting occurs and no blip is observed; the variation using a hole in the crucible circumvents this problem; (iii) the data acquisition rate must be high ($<0.5\,s$ per point) to record the baseline blip; and (iv) placing the hook on the platinum wire is difficult. The drop method is not suited to all instrument designs and also cutting a hole in the crucible alters the environment of the standard reference material compared with the sample whose crucible has no hole. Drop methods do have the advantage that well characterized suitable reference materials are easily obtained (Appendix 2.1).

When thermobalances are sold as simultaneous TG–DTA (TG–DSC) apparatus, temperature calibration is most conveniently carried out using the techniques for DSC calibration described in Chapter 3. Temperature calibration, by whatever method, should be performed under conditions identical with those of the proposed experiment. A large proportion of the mass loss steps observed in TG occur over several tens of degrees and therefore highly accurate temperature calibration is not always essential.

4.4 Sample

The sample mass, volume and form greatly affect the characteristics of a TG curve. In considering the optimum mass of sample the following points should be noted: (i) endothermic and exothermic reactions occurring in the sample cause the true sample temperature to deviate significantly from the programmed temperature, and this deviation increases with increasing mass of the sample and can be as large as 50 K; (ii) the evolution of gases from the sample depends on the nature of the environment immediately surrounding individual particles, which is to some extent determined by the bulk of sample in the crucible; and (iii) thermal gradients are more pronounced for large sample masses, particularly in the case of polymers where thermal conductivities tend to be low. For these reasons, it is recommended to use as little sample as possible within the limits of resolution of the microbalance. The homogeneity of a sample can sometimes limit how little sample can be used. This is particularly true for polymer blends. Owing to the influence of bouyancy effects (Section 4.5), the sample volume should also be kept to a minimum.

Powdered samples, of small particulate size, have the ideal form for TG studies. However, in polymer science samples are often in the form of films, fibres, sheets, pellets, granules or blocks. The TG curves recorded for a PMMA sample with various physical forms are presented in Figure 4.12A. The shapes and residue rates of these curves are strikingly different. The temperature at which the rate of mass loss is a maximum is also affected by the physical form of the polymer. DTA curves recorded simultaneously and presented in Figure 4.12B reveal large differences in the sample behaviour. A variety of techniques can be employed to obtain a sample form suitable for TG analysis and there is considerable room for individual initiative. The method of preparation should not alter the microscopic structure of the sample and should be reproducible. The sample may be cut using a microtome. This technique has the advantage of producing sections of predetermined and uniform thickness. Occasionally this method requires freezing of the sample before sectioning and in this case the effect of freezing on the sample characteristics should be verified. A blade can also be used to produce thin slices of sample. Both methods suffer from the drawback that a large shearing force is applied to the sample surface during slicing, which can alter the sample morphology. Cork borers and punches can also be used to produce disc-shaped samples with reproducible dimensions. In preparing fibres for TG analysis the greatest problem is reproducing the surface-to-volume ratio of the sample. Careful grinding of the fibres is one solution so that the sample may be packed in an efficient and reproducible manner. Grinding should be carried out with

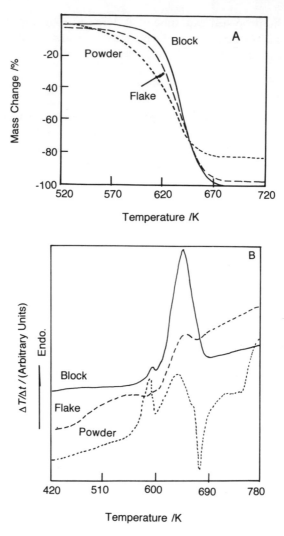

Figure 4.12. (A) TG curves for PMMA with various physical forms of sample. ————, Block; – –, flake, - - -, powder. (B) DTA curves recorded simultaneously. Experimental conditions: amount of sample 5 mg; sample, heating rate 5 K/min; dry N_2 purge gas 20 ml/min

minimal force and slowly enough so that thermal effects due to friction or reorientation are avoided. Alternatively, the fibres can be cut into small lengths using a blade or guillotine. Liquid samples should be dispensed using a microsyringe.

Figure 4.13A shows the TG curves recorded for a powdered PMMA sample in a densely packed and a loosely packed configuration. The shapes of these

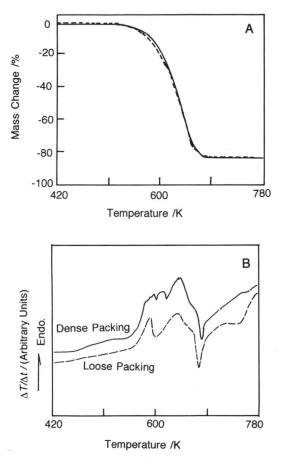

Figure 4.13. (A) Effect of packing density on the TG curve of powdered PMMA. Solid line, densely packed; dashed line, loosely packed. (B) DTA curves recorded simultaneously. Experimental conditions: amount of sample, 5 mg; heating rate, 5 K/min; dry N_2 purge gas flow rate, 20 ml/min

curves are similar, but the temperature at which the rate of mass loss is a maximum is increased by 2 K for the densely packed sample. However, simultaneous DTA measurements, presented in Figure 4.13B, reveal that despite the similarity of the TG curves the decomposition processes occurring in the sample are greatly influenced by the packing density. The packing density should be as uniform as possible. In a loosely packed sample thermal conductivity is low owing to the poor conductivity of the air between the particles. Dense packing, on the other hand, can inhibit the diffusion of evolved gases through the sample and significantly alter the observed decomposition reaction kinetics. Even for powdered samples variations in the TG curves are observed with changes in particle size. Smaller particles have the best surface-to-volume ratio and at any given temperature the extent of decomposition is greater than for samples of larger particle size. The smaller the particle size, the greater is the extent to which equilibrium is reached, all other conditions being equal. It is recommended to try several different sample preparation procedures in order to identify and circumvent artifacts, before standardizing the preparation method.

4.5 Atmosphere

TG analysis can be performed under a variety of atmospheric conditions, which include high pressure (up to 300 MPa), vacuum (down to 10^{-3} Pa) and atmospheric pressure in the presence of inert, corrosive, oxidizing or reducing gases. High-pressure work requires a metal casing and high-pressure sealing around the sample housing. Working at reduced pressures typically requires quartz housing around the sample while diffusion pumps are used for evacuating the chamber. Atmospheres can be broadly divided into two classes, namely interactive and non-interactive. A non-interactive atmosphere such as helium gas is used to standardize the conditions of the experiment without affecting the sample. An interactive atmosphere such as oxygen can play a direct role in the reaction processes taking place in the sample. Atmospheres can be employed under static or flowing conditions, the latter being used to remove evolved gases from the vicinity of the sample. Gases employed in TG analysis include air, Ar, Cl_2, CO_2, H_2, HCN, H_2O, He, N_2, O_2 and SO_2. Before attempting to use Cl_2, HCN or SO_2 it is strongly recommended to consult with a safety officer to ensure the safe use and proper disposal of these harmful gases.

It is common to purge the air from the sample chamber owing to the relatively complex composition of the air and the fact that air contains a large proportion of highly interactive water vapour. The region of the microbalance mechanism should be constantly maintained under an inert, flowing atmosphere such as dry He or dry N_2, even when the thermobalance is not in use. A flow rate of 30–40 ml/min is recommended during operation and 1–2 ml/min after use. The water content of the purge gas should be less than 0.001% (g/g). The sample chamber can be purged using the same gas as in the microbalance housing or another gas. The flow rate depends on the size of the sample, but should be less than that used in the microbalance housing to avoid contamination of the latter. A flow rate of 15–25 ml/min is recommended for samples between 2 and 10 mg.

Variations in the behaviour of the atmosphere under different experimental conditions can result in erroneous or misleading data. The density of a gas is a function of temperature, pressure and the nature of the gas. A change in the density of the purge gas during heating can result in an apparent mass change of the sample. This is called the bouyancy effect and samples which appreciably change their volume in addition to their mass during heating are particularly susceptible to bouyancy effects. Owing to the non-uniform heating of the furnace chamber, temperature gradients are produced along the crucible support which at low pressures cause streaming of the purge gas in the direction of the temperature gradient producing spurious apparent mass changes. The simplest way to correct for bouyancy and streaming effects is to run an inert sample, of similar dimensions to the actual sample, under identical experimental conditions, and subtract this background TG curve from the experimental curve. When working at high temperature and high pressure, the large thermal gradients present can give rise to convection effects which can cause apparent mass changes to be recorded. The introduction of baffles reduces convection effects and consultation with the manufacturer is recommended to choose the best configuration given the proposed experimental conditions.

Under vacuum conditions (< 0.1 Pa) with a gas-evolving sample, particularly at high scanning rates, a mass increase may be observed. This apparent mass change is due to re-impacting of gas molecules on the crucible and can usually be eliminated by reducing the heating rate or increasing the pumping cross-section. Volatile products released during heating may condense on the crucible support which is at a lower temperature and give rise to an apparent sample mass change. One method to check for condensation is to weigh the crucible and support before and after the experiment to determine if condensation has occurred. Condensation effects can be avoided altogether by sheathing the crucible support, for example with a ceramic sleeve. The sample chamber housing is frequently made of glass and a considerable amount of electrostatic charge can build up on the glass surface attracting the crucible and interfering with the balance mechanism. This can be avoided by applying an anti-static spray to the housing or by swabbing (not rubbing) the housing with a damp cloth made from natural fabric.

TG experiments may be conducted in the presence of a self-generated atmosphere. This is achieved by placing the sample in a crucible of small vapour volume with a small opening to the atmosphere of the furnace and with the exception of the gas initially present in the crucible decomposition occurs in the presence of the volatile or gaseous decomposition products. A variety of crucible shapes and designs are available for such experiments (Figure 4.4, IX and X). The operational difficulties of self-generated atmospheres are as follows: (i) fluctuations in the nature and composition of the self-generated atmosphere can cause bouyancy effects which are difficult to correct; (ii) the crucibles used tend to have a large thermal mass and therefore the sample temperature is less well defined; and (iii) secondary reactions between the solid and gaseous phases can complicate interpretation of results. The advantages claimed for self-generated atmosphere studies include[3]: (i) a narrow reaction interval and better resolution of overlapping reactions; (ii) the observed initial decomposition temperature is more closely related to the equilibrium decomposition temperature; and (iii) improved resolution of irreversible decompositions.

4.6 Heating Rate

The heating rate has a strong influence on the shape of a TG curve. The most visible effect is on the procedural decomposition temperatures T_i and T_f. For a one-step endothermic decomposition reaction the following is observed: (i) $(T_f)_h > (T_f)_l$, (ii) $(T_i)_h > (T_i)_l$ and $(T_f - T_i)_h > (T_f - T_i)_l$, where the subscripts h and l denote high and low heating rates, respectively. A sample will begin to decompose when the vapour pressure of the gaseous products exceeds the ambient partial pressure. When the temperature is reached at which this condition is satisfied, decomposition will occur if the product gases can freely diffuse from the sample. In the case of dense packing and/or a high heating rate, such free diffusion is inhibited and the decomposition temperature is increased. At low heating rates the sample temperature is more uniform and diffusion of product gases can occur within the sample lowering the decompositional temperature. Furthermore, because the sample is decomposing in an atmosphere which is more constant than at a higher heating rate, the decomposition reaction will be completed within a narrower temperature interval. The disparity between the true sample temperature and the programmed temperature is increased at high heating rates. Also, the extent of sample decomposition is greater at a lower heating rate and where multiple reactions occur the resolution of individual reactions is reduced at high heating rates. Figure 4.14A presents the TG curves of powdered poly(vinyl chloride) (PVC) samples recorded at 2 and 20 K/min. These curves are clearly not superimposable. The residue rates and the physical form of the residues remaining in the crucible are different. The decomposition process occurring in the region of 575 K is more clearly resolved at the slower heating rate. Simultaneous DTA measurements result in very different curves (Figure 4.14B). A heating rate in the range 5–10 K/min is recommended.

A sample can also be observed under isothermal conditions where the mass change is recorded as a function of time at a predetermined temperature. A third method of TG analysis is the jump method[4]. In this case the sample is held at a fixed temperature for a period of time until the temperature is discontinuously changed (or jumped), where again the mass change is observed as a function of time. The application of the jump method to the study of reaction rate kinetics is discussed in Section 5.3.3. If the chemical reaction under investigation proceeds slowly then the linear heating programme may be replaced by a stepwise programme so that the experimental conditions become quasi-isothermal[5]. The effect of altering the heating programme on the shape of a TG curve of a slow, single-step decomposition reaction is shown in Figure 4.15.

A more complex system of temperature and atmosphere control used in the study of decomposition rates is quasi-isothermal and quasi-isobaric thermogravimetry[6]. In this method a heating programme linearly increases the sample temperature until a mass change is detected. On detection of a mass

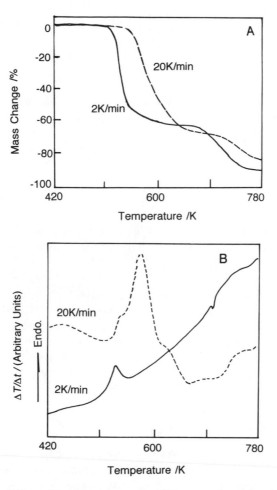

Figure 4.14 (A).   Effect of heating rate on the TG curve of powdered PVC. Solid line, 2 K/min; dashed line, 20 K/min. (B) DTA curves recorded simultaneously. Experimental conditions: amount of sample, 5 mg; dry N_2 purge gas flow rate, 20 mL/min

change the temperature programme is automatically altered so that decomposition takes place at a predetermined rate (typically 0.5 mg/min), thereby establishing quasi-isothermal conditions for the sample. Effects due to slow heat transfer through the sample on the rate of reaction are minimized by this method. Also, the large temperature fluctuations due to the endothermic or exothermic nature of the reaction process are reduced. When the sample mass becomes constant following completion of the reaction the sample is again heated in a linear manner until the next mass loss step is detected. Quasi-isobaric conditions are maintained by elaborate crucible designs (Figure 4.4, X) used to produce a self-generated atmosphere, where the partial pressure of

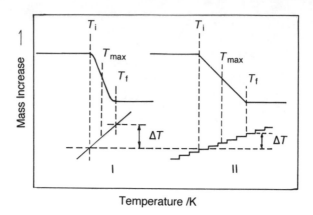

Temperature /K

Figure 4.15. Schematic representation of the TG curve of a single stage decomposition process in the presence of (I) a linear and (II) a stepwise temperature programme

the gaseous decomposition products in contact with the solid phase is approximately constant. The object of quasi-isothermal and quasi-isobaric thermogravimetry is to make the shape of the TG curve independent of effects due to the experimental conditions and therefore more amenable to kinetic analysis. To facilitate estimation of reaction rate kinetic parameters, hyperbolic heating profiles are sometimes employed of the form $1/T = r + st$, where r and s are constants (Section 5.3).

4.7 Classification of TG Curves

A scheme for classifying TG curves has been proposed[7] where the curves are classified according to their shape into one of seven categories. Each category is schematically represented in Figure 4.16. Type A curves show no mass change over the entire temperature range of the experiment and the only information gleaned from the curve is that the decomposition temperature of the material is greater than the maximum temperature, under the experimental conditions. Other TA techniques such as DSC can be used to investigate whether non mass-changing processes have occurred. A large initial mass loss followed by a mass plateau characterize type B curves. Evaporation of volatile components used during polymerization, drying and desorption processes give rise to such curves. Where a non-interacting atmosphere is present the second run of a type B sample will result in a type A curve.

The third category, type C, is a single-stage decomposition reaction where the procedural decomposition temperatures (T_i and T_f) are used to characterized the curve. Multi-stage decomposition processes where the reaction steps are clearly resolved comprise type D curves. This is in contrast to type E where the individual reaction steps are not well resolved. In the case of type E the

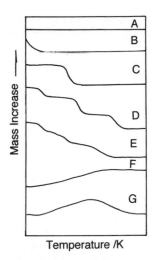

Figure 4.16. Classification scheme for TG curves. (Reproduced by permission of Elsevier Science Publishers from Duval, *Inorganic Thermogravimentric Analysis*, 1963)

DTG curve (Section 4.9) is often preferred as the characteristic temperatures may be determined more accurately. In the presence of an interacting atmosphere a mass increase may be observed giving rise to a type F curve. Surface oxidation reactions belong to this category. The final category, type G, is not frequently encountered and could be caused, for example, by surface oxidation followed by decomposition of the reaction products.

4.8 Calculation of Mass Change Rates

The International Standards Organization (ISO) recommends the following nomenclature and procedures for determining the characteristic temperatures and masses of a TG curve. Using a single-stage mass decrease TG curve as an example ISO defines the following: (i) the starting point, **A**, as the intersection between the starting mass line and the maximum gradient tangent to the TG curve; (ii) the end point, **B**, as the point of intersection of the minimum gradient tangent following completion of the mass decrease step and the maximum gradient tangent to the TG curve; and (iii) the intersection between the TG curve and a line drawn parallel to the horizontal axis through the mid-point of **A** and **B** is defined as the mid-point **C**. The masses m_s and m_B and the temperatures T_A, T_B and T_C, associated with **A**, **B** and **C** are estimated as in Figure 4.17.

In the case of a multi-stage decomposition, the characteristic masses and temperatures are estimated as in Figure 4.18. Where the thermobalance does not record a region of constant mass between successive decomposition stages, two minimum tangents should be drawn, the first following completion of the primary stage and the second before the secondary stage begins.

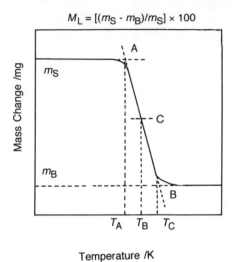

Figure 4.17. Characteristic temperatures and masses for a single stage mass loss curve. The rate of mass loss, M_L, is given by the equation

In the case of increasing mass, Figure 4.19 illustrates the procedure for calculating the maximum mass attained and the rate of increase in mass. The residue rate $R = (m_f/m_s) \times 100$, where m_f and m_s are the sample mass before heating and following completion of the final stage, respectively.

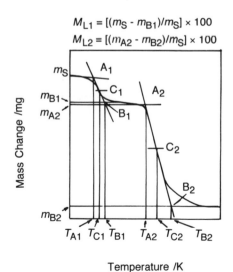

Figure 4.18. Characteristic temperatures and masses for a multi-stage mass loss curve. The rate of mass loss at each stage, M_{Li}, is given by the equation

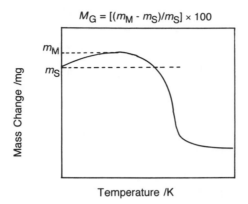

$$M_G = [(m_M - m_S)/m_S] \times 100$$

Figure 4.19. Characteristic masses for a single-stage mass gain curve. The rate of mass gain, M_G, is given by the equation

4.9 Derivative Thermogravimetry (DTG)

Overlapping reactions are sometimes difficult to resolve and in some cases resolution may be improved by hindering the escape of evolved gases from the sample by placing a loosely fitting lid on the crucible, changing the packing or form of the sample, choosing a different crucible or by varying the heating rate. Altering the experimental conditions may change the relative rates of the overlapping reactions and lead to better resolution. Often it is undesirable to alter optimized experimental conditions and in this case overlapping reactions may be more clearly resolved by plotting the derivative TG curve. In DTG the mass change with respect to temperature (dm/dT) is plotted against temperature or time (Figure 4.20A). A point of inflection in the mass change step becomes a minimum in the derivative curve and for an interval of constant mass dm/dT is zero. A peak in the DTG curve occurs when the rate of mass change is a maximum. DTG peaks are characterized by the peak maximum temperature (T_{max}) and the peak onset temperature (T_e). Figure 4.20B shows how a DTG curve can be used to resolve overlapping reactions. The area under a DTG curve is proportional to the mass change and the height of the peak at any temperature gives the rate of mass change at that temperature. However, a DTG curve contains no more information than the original TG curve and therefore T_{max} and T_e are only as representative as T_i and T_f, and similarly have no absolute significance. DTG curves are frequently prefered when comparing results with DTA curves because of the visual similarity.

4.10 Intercomparison of TG and DTA

Like many experimental techniques, TG analysis does not give an unequivocal result for every physico-chemical phenomenon encountered. Intercomparison

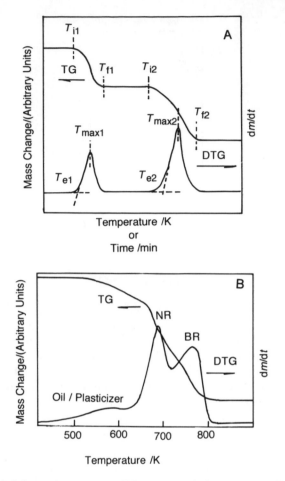

Figure 4.20. (A) Schematic two stage TG curve and the corresponding DTG curve. (B) Decomposition of a natural rubber–butadiene rubber blend is more clearly resolved into individual component decomposition reactions using the DTG curve. (Reproduced with permission from *Rubber Chemistry and Technology*, **48**, 661 (1975))

of data from different techniques is often required to interpret a result unambiguously. DTA is the TA technique most frequently employed for intercomparison with TG results. Data collected using simultaneous TG–DTA apparatus are preferred, but data from separate measurements can be compared with confidence provided that the experimental conditions are precisely controlled. A variety of simultaneous TG–DTA instrument designs are available. Figure 4.21 illustrates schematically the TG and DTA curves obtained for a variety of physico-chemical processes. Over the entire temperature range of an experiment several of these processes may occur in succession.

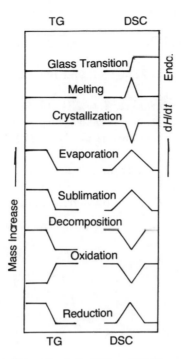

Figure 4.21. Comparison of the schematic TG and DSC curves recorded for a variety of physico-chemical processes

4.11 TG Reports

In this chapter, the influence of experimental conditions on the shape and characteristics of TG curves has been stressed. When preparing a report on a series of TG experiments along with a description of the results, the following experimental parameters should be noted:

- sample identification;
- form and dimensions of sample;
- mass of sample, initial and final;
- preconditioning of sample;
- crucible shape, size and material;
- type and position of sample thermocouple;
- atmosphere and gas flow rate;
- heating rate or isothermal temperature;
- reference material used for temperature calibration;
- type of instrument used.

With such a report results from different sources can be meaningfully compared and the cause of disparities determined.

4.12 References

[1] Norem, S.D., O'Neill, M.J. and Gray, A.P. *Thermochimica Acta* **1**, 29 (1970).
[2] McGhie, A.R., Chiu, J., Fair, P.G. and Blaine, R.L. *Thermochimica Acta* **67**, 241 (1983).
[3] Newkirk, A.E. *Thermochimica Acta* **2**, 1 (1971).
[4] Flynn, J.H. and Dickens, B. *Thermochimica Acta* **15**, 1 (1976).
[5] Oswald, H.R. and Wiedemann, H.G. *Journal of Thermal Analysis* **12**, 147 (1977).
[6] Arnold, M., Veress, G.E., Paulik, J. and Paulik, F. *Journal of Thermal Analysis* **17**, 507 (1979).
[7] Duval, C. *Inorganic Thermogravimetric Analysis*, Elsevier, Amsterdam, 1963.

5 APPLICATIONS OF THERMAL ANALYSIS

5.1 Temperature Measurement

A DSC melting peak is presented in Figure 5.1A, where the characteristic temperatures of the peak are also illustrated. The subscripts i, p and e refer to the initial, peak and end temperatures, respectively. The subscript m denotes melting. T'_{im}, the onset temperature, is the temperature at which the first deviation from linearity of the sample baseline is observed. The extrapolated onset temperature, T_{im}, is defined as the intersection between the tangent to the maximum rising slope of the peak and the extrapolated sample baseline. Generally the extrapolated sample baseline is taken as a straight line joining the points

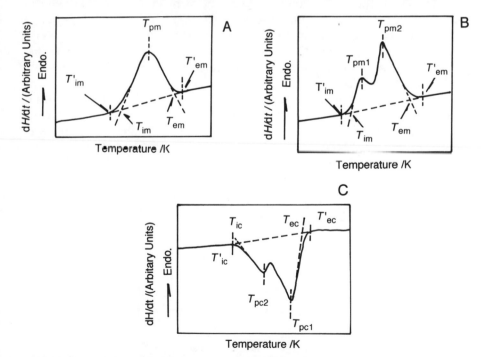

Figure 5.1 Schematic representation of DSC curves in the region of a phase transition and the characteristic temperatures. (A) Melting endotherm. (B) Melting endotherm with multiple peaks. (C) Crystallization exotherm with multiple peaks

of deviation from the sample baseline before (T'_{im}) and after (T'_{em}) the peak. This procedure is approximate and lacks thermodynamic justification. In the case of high-sensitivity DSC, where the recorded heats of transition are considerably smaller, the error introduced by this approximation becomes very significant and more representative formalisms are used to estimate the extrapolated sample baseline (Section 5.13). The melting peak temperature is denoted T_{pm}. T_{em}, the extrapolated end temperature, is given by the intersection of the tangent to the maximum falling slope of the peak and the extrapolated sample baseline. In the case of multiple peaks the characteristic temperatures are defined as in Figure 5.1B. The characteristic temperatures of crystallization peaks are similarly defined (Figure 5.1C), where the subscript c denotes crystallization.

The thermal history of a polymer has a strong influence on the characteristic melting and crystallization temperatures. When the effect of thermal history is of interest (for example, following thermal processing, annealing, ageing, curing) the sample should be examined without any further conditioning. Frequently it is desirable to erase the previous thermal history of a sample and in this case the following preliminary thermal cycle is recommended. The sample is placed in the DSC at ambient temperature and heated at 10 K/min to 30 K above the sample melting point, or a temperature high enough to erase the previous thermal conditioning. The sample is maintained at that temperature for approximately 10 min, but not long enough for decomposition or sublimation to occur, before cooling at the same rate to 50 K below the crystallization peak temperature. From this temperature the sample can be thermally cycled and the characteristic temperatures determined from the observed melting and crystallization peaks.

The glass transition temperature of a polymer can be determined using DSC, where the characteristic temperatures are defined as in Figure 5.2 and the subscript g refers to the glass transition. A preliminary thermal cycle similar to that cited above can be applied to polymers where T_g is of interest when the previous thermal history has been erased.

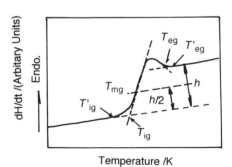

Figure 5.2.   Schematic representation of a DSC heating curve in the region of a glass transition and the characteristic temperatures

Not all characteristic temperatures can be reproduced with equal accuracy and the use of T_{im}, T_{pm}, T_{em} and T_{ic}, T_{pc}, T_{ec} is recommended when reporting the melting and crystallization peak characteristics, respectively. In the case of the glass transition only T_{ig} and T_{mg} can be accurately reproduced and should be used for comparison with data from other sources.

5.2 Enthalpy Measurement

The area of a DSC peak can be used to estimate the enthalpy of transition, ΔH. The thermal history of the sample must be considered when measuring ΔH. For example, annealing tends to improve the stability of the high-order structure of polymers and therefore affects the measured value of ΔH. A large variation in the reported values of ΔH can result from the difficulty in determining the points of deviation from the sample baseline (T'_i and T'_e), especially if pre-melting occurs or the natural width of the peak is large (Figure 5.3)[1]. This problem is compounded if the instrument baseline is curved. There are two principal methods for determining ΔH. The most common method is to calibrate the energy scale of the instrument with standard reference materials whose enthalpies of melting are well characterized and whose melting temperatures extend over a wide temperature range (Appendix 2.1). By calibration a certain peak area corresponds to a known value of ΔH and under identical experimental conditions, from the area of the sample peak, the enthalpy of transition can be estimated. The peak area can be measured using

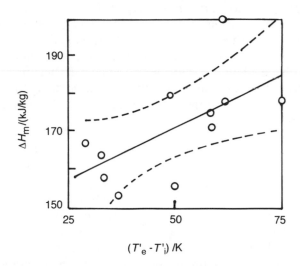

Figure 5.3. Relationship between the melting enthalpy (ΔH_m) and the difference between the onset and melting temperature ($T'_{im} - T'_{em}$) of polyethylene film. This plot is compiled from the data reported by 12 researchers. (Reproduced by permission of Elsevier Science Publishers from *Thermochimica Acta*, **138**, 327 (1989)

one of the following methods: (i) cut the peak from the chart recorder trace and weigh the piece of paper; (ii) measure the peak area with a planimeter (minimum peak area $5\,cm^2$) or (iii) select the software option which automatically calculates the peak area for an inputted T'_i and T'_e. If methods (i) or (ii) are used the peak area should be the average of five determinations expressed in cm^2 and the following equation used to calculate the enthalpy of melting (or crystallization):

$$\Delta H = \frac{ABT}{W} \frac{\Delta H_s W_s}{A_s B_s T_s} \tag{5.1}$$

where ΔH_s is the enthalpy of melting of the standard reference material (J/g), A and A_s are the peak area of the sample and standard (cm^2), W and W_s are the weight of the sample and standard (mg), T and T_s are the vertical axis sensitivity of the sample and standard (mW/cm) and B and B_s are the horizontal sensitivity of the sample and standard (cm/min), respectively.

An alternative method[2] of determining ΔH uses the fact that in power compensation DSC the proportionality constant between the transition peak area and ΔH is equivalent to the constant which relates the sample heat capacity and the sample baseline increment. By measuring the specific heat capacity of a standard sapphire sample, an empty sample vessel and the sample of interest, from the difference in the recorded DSC curves of the three experiments, ΔH for the sample transition can be calculated. The advantage of this method is that sapphire of high purity and stability, whose specific heat capacity is very accurately known, is readily available. Only one standard material (sapphire) is necessary irrespective of the sample transition temperature. The linear extrapolation of the sample baseline to determine ΔH has no thermodynamic basis, whereas the method of extrapolation of the specific heat capacity in estimating ΔH is thermodynamically reasonable. The major drawbacks of this method are that the instrument baseline must be very flat and the experimental conditions are more stringent than for the previous method. Also, additional computer software and hardware are required to perform the calculation.

5.3 Reaction Rate Kinetics

TA instruments can be used to investigate the reaction rate kinetics of a broad range of materials, including polymers. The two basic approaches to determining reaction rate kinetic parameters are isothermal methods and non-isothermal (dynamic) methods. In an isothermal experiment the sample is quickly brought to a predetermined temperature where the TA instrument monitors the behaviour of the system as a function of time. Non-isothermal methods record the response of the sample as it is scanned, usually at a constant rate although hyperbolic temperature programmes are sometimes used. In the case of polymers the sample morphology and structure can change during the preliminary heating stage of an isothermal experiment. These initial changes in structure are for the most part uncontrolled and their occurrence greatly complicates analysis of the isothermal curve. Non-isothermal analysis is generally preferred for the following reasons: (i) dynamic experiments are quicker and the results widely considered easier to interpret; (ii) the reaction process can be followed over a wide temperature range; (iii) several reaction steps can be observed in a single experiment; and (iv) a number of methods of data evaluation are available. In addition, it is commonly (but incorrectly) believed that a single dynamic curve is equivalent to a large number of comparable isothermal curves and that the theories developed for estimating isothermal reaction rate kinetic parameters can be applied to non-isothermal data.

Reacting systems are divided into two classes, homogeneous and heterogeneous. In homogeneous systems the rate of reaction is determined by measuring isothermally the decrease in concentration of the reactant or the increase in concentration of the product. The reaction is usually considered to be activated where the reaction occurs when a particle of reactant attains sufficient energy to overcome the potential energy barrier impeding the reaction. The rate of reaction is given by

$$\frac{d[X]}{dt} = -kf([X]) \tag{5.2}$$

where [X] is the concentration of reactant and k the rate coefficient, which is given by the Arrhenius equation:

$$k = A\exp(-E/RT) \tag{5.3}$$

The units of k are s^{-1}; A (s^{-1}) is the pre-exponential factor (sometimes called the frequency factor), E (kJ/mol) is the activation energy, T (K) is the temperature of the sample and R (J/mol K) is the gas constant. A temperature dependence for A can be expressed by replacing A with $A' = A_n T^n$. The form of $f([X])$ depends on the presumed nature of the reaction process. By monitoring the change in reactant concentration as a function of time, at various temperatures, A and E can be determined graphically.

In the above treatment, a number of explicit assumptions are made concerning both the sample and the experimental conditions. These assumptions can be summarized as follows: (i) the sample is entirely homogeneous; (ii) the reaction process is activated and no geometric or interfacial effects are present; (iii) the reaction proceeds by elememtary sequential steps and no overlapping or parallel reactions occur; (iv) the temperature of the sample is uniform and no thermal gradients exist in the sample; (v) there are no compositional gradients present; (vi) where gas is evolved there is no impedance to gas diffusion and there are no partial pressure gradients; and (vii) the preliminary heating stage has no effect on the sample, irrespective of the heating rate.

Almost all solid-state reacting systems are heterogeneous in nature. The heterogeneity can be compositional and/or structural. This is particularly true for polymer systems where for a given homopolymer distributions in molecular weight, tacticity, cross-linking density, degree of orientation, functional group concentration and the presence of plasticizers, unreacted monomer, residual catalyst and stabilizers render the system chemically and physically heterogeneous. Heterogeneity is increased in the case of copolymers or polymer blends by the addition of extra components to the system. TA of polymers is generally carried out using non-isothermal methods and under these conditions it is difficult to measure the concentration of any given reacting species without altering the reaction kinetics. The progress of the reaction is followed by monitoring one of the following as a function of temperature:

(i) TG—the fraction of starting material consumed, α:

$$\alpha = \frac{m_s - m}{m_s - m_f} \tag{5.4}$$

where m_s, m and m_f are the initial, actual and final mass of the sample, respectively;

(ii) DSC—the rate of change of heat evolved or consumed over the reaction interval, dH/dt;

(iii) DTA—the difference temperature between the sample and reference over the reaction interval, $\Delta T/\Delta t$.

Estimates of reaction rate kinetic parameters should be based on data acquired from TG curves because the mass change demonstrates most reliably the progress of reactant conversion and there exists a strict numerical correlation between the measured property (α) and the progress of a reaction. This is not the case for either DSC or DTA.

The expression for the rate of reaction of a heterogeneous system has the following general form:

$$\frac{d\alpha(t)}{dt} = k(T)f[\alpha(t)]h(\alpha, T) \tag{5.5}$$

The function $h(\alpha, T)$ is normally taken to be equal to unity, with little justification for this approximation. Despite the heterogeneous nature of the system and the inherent assumptions of the Arrhenius equation the rate coefficient, k, is generally expressed as in equation 5.3. The various functional forms of $f[\alpha(t)]$ are presented in Table 5.1. In its most commonly presumed form for solid-state reactions $f[\alpha(t)] = (1 - \alpha)^n$, where n is the reaction order and is assumed to remain constant for the duration of the reaction. Equation 5.5 is usually written as

$$\frac{d\alpha(t)}{dt} = A\exp(-E/RT)(1 - \alpha)^n \tag{5.6}$$

Table 5.1. Commonly used functional forms of $f[\alpha(t)]$

Designation	$f[\alpha(t)]$	Description of reaction process
A2	$2(1-\alpha)[-\ln(1-\alpha)]^{\frac{1}{2}}$	Random nucleation, Avrami–Erofe'ev equation
A3	$3(1-\alpha)[-\ln(1-\alpha)]^{\frac{2}{3}}$	Random nucleation, Avrami–Erofe'ev equation
A4	$4(1-\alpha)[-\ln(1-\alpha)]^{\frac{3}{4}}$	Random nucleation, Avrami–Erofe'ev equation
D1	$1/2\alpha$	One-dimensional diffusion
D2	$[-\ln(1-\alpha)^{\frac{1}{3}}]^{-1}$	Two-dimensional diffusion
D3	$3/2(1-\alpha)^{\frac{2}{3}}[1-(1-\alpha)^{\frac{1}{3}}]^{-1}$	Three-dimensional diffusion
F1	$1-\alpha$	First-order
F2	$(1-\alpha)^2$	Second-order
F3	$(1-\alpha)^3$	Third-order
R2	$2(1-\alpha)^{\frac{1}{2}}$	Cylindrical phase boundary
R3	$3(1-\alpha)^{\frac{2}{3}}$	Spherical phase boundary

Under dynamic conditions the sample temperature is presumed to be a function of time only and equal to the furnace temperature $[dt = \psi(T)dT]$. Substituting into equation 5.6 we obtain

$$\frac{d\alpha}{(1-\alpha)^n} = \psi(T)A\exp(-E/RT)dT \qquad (5.7)$$

Ideally, integration of equation 5.7 should reproduce the TG curve of a given sample. Integration can be easily performed if a hyperbolic temperature programme of the form $1/T = r - st$ is used, where r and s are constants. For a linear heating programme with constant heating rate, ϕ, equation 5.7 becomes

$$\frac{d\alpha}{(1-\alpha)^n} = \frac{A}{\phi}\exp(-E/RT)dT \qquad (5.8)$$

The right-hand side of equation 5.8 cannot be integrated and an abundance of methods for estimating non-isothermal kinetic parameters deal with this problem so that values of A, E and n can be calculated from a single curve or from a series of curves recorded at different heating rates. These methods are based on the hypothesis that A, E and n uniquely characterize a given reaction irrespective of the experimental conditions.

The schemes for dealing with equation 5.8 are normally classified as either differential or integral methods and an introduction to the most frequently encountered procedures is presented next.

5.3.1 DIFFERENTIAL METHODS

A method that is widely employed to calculate rate parameters is that of Freeman and Carroll[3], where equation 5.6 is re-written in logarithmic form as

$$\ln(d\alpha/dt) = n[\ln(1 - \alpha)] + \ln A - E/R(1/T) \qquad (5.9)$$

Differentiating with respect to $\ln(1 - \alpha)$:

$$\frac{d[\ln(d\alpha/dt)]}{d[\ln(1 - \alpha)]} = n - E/R \; \frac{d(1/T)}{d[\ln(1 - \alpha)]} \qquad (5.10)$$

Therefore, a plot of

$$\frac{d[\ln(d\alpha/dt)]}{d[\ln(1 - \alpha)]} \text{ versus } \frac{d(1/T)}{d[\ln(1 - \alpha)]}$$

results in a straight line of slope $-E/R$ and intercept n. The order of reaction and the activation energy are calculated from a single experimental curve. A problem in calculating n by this method is that because the slope of the best-fit line to the data points has a very large absolute value, a small error in the estimation of the slope results in considerable uncertainty in the value of n. As a result, it is often impossible to distinguish with confidence between the various proposed reaction mechanisms using this procedure. An improvement to this method[4] only considers the points on the TG curve where the reaction rate is a maximum, resulting in a considerably smaller error in estimating n. The kinetic parameters estimated using the Freeman and Carroll method[3] show a strong dependence on the sample mass and the heating rate. This method has been extended to DSC analysis of reactions. However, the reported kinetic parameters estimated from different parts of the DSC curve are significantly different. It has been proposed[5] that no physical meaning can be attributed to rate parameters obtained using this method.

Originally proposed as a method for calculating kinetic parameters for reactions of the type solid $\rightarrow$ solid + gas from DTA experiments, Kissinger's method[6] also assumes that the reaction rate is described by equation 5.6. The most important additional assumption of this method is that the maximum in the DTA curve occurs at the same temperature as the maximum reaction rate. The reaction is further assumed to proceed at a rate which varies with temperature and therefore the position of the DTA peak is a function of heating rate. From the variation in the peak temperature with heating rate E can be calculated for any value of n. The maximum reaction rate occurs when $d/dt(d\alpha/dt) = 0$. From equation 5.6 it follows that

$$\frac{E\phi}{RT_m^2} = An(1 - \alpha)_m^{n-1}\exp(-E/RT_m) \qquad (5.11)$$

where T_m is the peak maximum temperature. By substituting an approximate solution to equation 5.6 into equation 5.11 and differentiating, it can be shown that

$$\frac{d(\ln\phi/T_m^2)}{d(1/T)} = -E/R \qquad (5.12)$$

Kissinger's method assumes that

$$d\alpha/dt = (\delta\alpha/\delta t)_T + (\delta\alpha/\delta T)_t dT/dt \tag{5.13}$$

and therefore the size and shape of the sample holder and the dilution of the sample do not affect the reaction rate. Under the same assumptions this method has been extended to include reactions which follow Avrami's law[7]. In this extension $\ln[\phi/(T_m - T_0)]$ is plotted against $1/T$, where T_0 is the starting temperature of the experiment. E is calculated from the slope of the best-fit line and n from the shape of the DTA curve. A generalized form of the Kissinger method has also been proposed[8] which includes a broad range of solid-state reaction mechanisms and a temperature dependence for A.

The method of Borchardt and Daniels[9] also uses the area of the DTA curve to calculate kinetic parameters. The additional assumptions made in deriving this method are pertinent to liquid systems, but extremely difficult to satisfy in solids. Despite the authors' stipulation to this effect, this method is frequently applied to the analysis of solid-state reactions. The additional assumptions include the following: (i) heat is transferred to the system by conduction alone; (ii) the heat capacity and heat transfer coefficients of the sample and reference are equal, and are independent of temperature over the reaction interval; (iii) the heat evolved is proportional to the number of moles reacting during a given time interval; and (iv) the volume of the sample does not change appreciably during the reaction. Under these assumptions the heat evolved by the reaction is given by

$$\Delta H = KP \tag{5.14}$$

where K is the heat transfer coefficient of the sample and P is the total area under the DTA curve. The expression for the reaction rate constant of an nth-order reaction is written as

$$k = -V^{n-1}(dx/dt)/x \tag{5.15}$$

where V is the sample volume and x the number of moles of reactant. If the correct value of n is chosen, a plot of $\ln k$ against $1/T$ will yield a straight line from which A and E can be estimated. This method is insensitive to errors because both the heat capacity and DTA cell constant cancel out in the derivation. A similar method[10] uses the area of the DSC curve to calculate E. This method is based on the relationship

$$k = -(PV/N_0)^{n-1}(dH/dt)/(P - a)^n \tag{5.16}$$

where P is the total area of the DSC curve, a is the area of the curve up to time t, V is the volume of the sample and N_0 the number of moles of reactant. The functional form $f[\alpha] = (1 - \alpha)^n$ is assumed and from a plot of $\ln k$ against $1/T$, E can be estimated.

5.3.2 INTEGRAL METHODS

Doyle[11] introduced a procedure for deriving kinetic data from a TG curve based on the assumption that a single non-isothermal TG curve is equivalent to a large number of comparable isothermal curves. Realizing the arbitrary nature of this assumption, the author treats the Arrhenius equation as empirical and recognizes the potential triviality of kinetic parameters derived using this relation. In the derivation equation 5.5 is re-written in an approximate form:

$$d\alpha/dt = kf(\alpha) \tag{5.17}$$

At a constant heating rate ϕ, substituting the Arrhenius equation for k and integrating, equation 5.17 becomes

$$\int_{\alpha_0}^{\alpha} \frac{d\alpha}{f(\alpha)} = g(\alpha) = -(A/\phi)\int_{T_0}^{T} \exp(-E/RT)dT \tag{5.18}$$

The function

$$p(E/RT) = (R/E)\int_{T_0}^{T} \exp(-E/RT)dT \tag{5.19}$$

is introduced and values of $p(E/RT)$ are calculated for the normal range of experimental values $10 \leqslant E/RT \leqslant 30$. When $E/RT \geqslant 20$ a linear approximation is made:

$$\log p(E/RT) \approx -2.315 - 0.4567(E/RT) \tag{5.20}$$

This method was further simplified by Ozawa[12] and applied to the random degradation of polymers where the proportion of sample remaining is defined in terms of the fractional number of broken bonds. Random degradation is observed in certain polymers where main-chain scission occurs at random points with equal probability. This method assumes that the degree of conversion is constant at the DTG peak temperature, for all heating rates. For a given fractional mass the left-hand side of equation 5.18 is constant and therefore

$$(AE/\phi_1 R)p(E/RT_1) = (AE/\phi_2 R)p(E/RT_2) = \ldots \tag{5.21}$$

Substituting Doyle's approximation:

$$-\log\phi_1 - 0.4567(E/RT_1) = -\log\phi_2 - 0.4567(E/RT_2) = \ldots \tag{5.22}$$

A plot of $\ln\phi$ against $1/T$, at a given fractional mass, yields a straight line from whose slope E can be calculated. Further, if the TG curves are plotted as fractional mass against $1/T$ at various heating rates the curves may be superimposed on each other by a lateral shift which is proportional to $\ln\phi$. A master curve is thereby obtained. If the TG curves cannot be superimposed in this manner, the Ozawa method is not applicable. In practice, the degree of conversion at the DTG peak temperature varies with ϕ.

A linearized approximation to equation 5.18 was proposed by Kassman[13] to evaluate reaction rate kinetic parameters. A reference temperature is chosen which is based on the initial and final temperatures of the experiment. A solution of the form $\ln g(\alpha) = \ln R + S\theta(T)$ is proposed, where R and S are constants and $\theta(T)$ has the form

$$[(T_r/T)\exp(1 - T_r/T)]^2 \approx 1 \tag{5.23}$$

T_r is the reference temperature and is given by

$$T_r = (T_1 T_2)^{\frac{1}{2}}\{1 - [(T_1 T_2)^{\frac{1}{2}}/S]\} \tag{5.24}$$

where T_1 and T_2 are the initial and final temperatures, respectively. From a plot of $\ln g(\alpha)$ against $1/T$, A and E can be estimated. The method proposed by Coats and Redfern[14] uses an asymptotic expansion to find an approximate solution to equation 5.18:

$$g(\alpha) = (ART^2/\phi E)(1 - 2RT/E)\exp(-E/RT) \tag{5.25}$$

$2RT/E \ll 1$ and is usually neglected. For an nth-order reaction mechanism a plot of

$$\frac{-\ln[1 - (1 - \alpha)^{1-n}]}{T^2(1 - n)}$$

for $n = 0$, $1/2$, $2/3$ or

$$\frac{-\ln[-\ln(1 - \alpha)]}{T^2}$$

for $n = 1$ against $1/T$ results in a straight line from which E is calculated.

5.3.3 JUMP METHOD

The jump method of Flynn and Dickens[15] is designed to circumvent the difficulties in estimating kinetic parameters due to the sample history. The rate forcing variable (temperature, pressure, gaseous composition, gas flow rate) of a TG experiment is changed discontinuously and the effect on the mass change recorded. The time constant of the system to return to quasi-equilibrium following the jump has three primary components: (i) the time constant of the instrument, (ii) the time constant of the sample and crucible and (iii) the response time of the reaction rate kinetic parameters to the new quasi-equilibrium conditions. It is suggested that measurement of the response time of the kinetic parameters should yield information concerning the intermediate stages of the reaction. The advantages claimed for the jump method are that the effects of sample history on the estimated parameters are reduced by confining these effects to the two short periods of extrapolation before and after the jump.

Also, the necessity of guessing the functional form of $f(\alpha)$ is avoided by obtaining reaction rate information corresponding to two (or more) values of the jumped variable at a single extent of reaction. There are several difficulties associated with this procedure. Discontinuous jumping of the experimental parameters is not easily achieved without a significant reduction in the resolution of the instrument and interpretation of the TG curve in the presence of jumps is not straightforward owing to the cumulative nature of the observed response time constant. Furthermore, it is difficult to choose the appropriate size of the jump and estimation of E from temperature jumps does not diminish the effect of the thermal history on the sample up to the jump.

5.3.4 ISOTHERMAL CRYSTALLIZATION OF POLYMERS

Isothermal crystallization of polymers is normally performed by cooling the sample at the fastest cooling rate possible from an (assumed) isotropic melt to a predetermined temperature where the crystallization process is monitored as a function of time. Coefficients similar to the reaction rate kinetic parameters have been developed to describe the isothermal crystallization of polymers based on Avrami's equation:

$$\ln \theta = -kt^n \tag{5.26}$$

where θ is a measure of the degree of crystallinity, k the crystallization rate constant, t the time and n a coefficient which depends on the model of nucleation (Table 5.2). A two-state model of the polymer structure is implicitly assumed, where the polymer is composed of distinct amorphous and crystalline regions. The densities of these regions are assumed to remain constant with time and the influence of thermal diffusion is neglected as a uniform temperature is assumed. The presence of compositional gradients and the effects of molecular mass are similarly ignored. The rate of nucleation is assumed to be constant over the entire crystallization interval and the phase transition goes to completion.

Table 5.2. Avrami index n for various nucleation processes

Homogeneous nucleation		Inhomogeneous nucleation: linear growth	Comment
Linear growth	Diffusion-controlled growth		
2	3/2	$1 \leqslant n \leqslant 2$	One-dimensional
3	2	$2 \leqslant n \leqslant 3$	Two-dimensional
4	5/2	$3 \leqslant n \leqslant 4$	Three-dimensional

From DSC data the degree of crystallization is estimated by plotting the fractional area of the isothermal crystallization curve as a function of time. The rate of nucleation as a function of crystallization temperature is shown schematically in Figure 5.4. The isothermal crystallization temperature should be choosen so that the crystallization proceeds at a rate which can be accurately measured. A practical difficulty is to determine the true zero time (t_0) from which the isothermal crystallization begins. The zero time cannot be determined exactly because the DSC is unstable following quenching. When using a quantitative DTA the zero time should be chosen as the earliest point where the sample baseline is flat, indicating that stable isothermal conditions have been reached (Figure 5.5A). If the isothermal crystallization temperature is chosen so that crystallization begins several minutes after the isothermal temperature has been attained, then the error in estimating the zero time introduced by adopting this convention is minimized. Users of power compensation-type DSCs should employ the same convention when a computer is used to process the data. However, when the isothermal curve is recorded on chart paper the zero time can be defined as the position on the chart when the control light comes on indicating that the difference between the programmed and sample temperatures is < 0.01 K (Figure 5.5B). The convention adopted to determine the zero time should be clearly stated when reporting isothermal crystallization results.

Another form of isothermal crystallization is observed with polymers which exhibit cold crystallization. In this case the sample is heated from the glassy state at the fastest rate possible to a temperature in the vicinity of the cold crystallization exotherm. The zero time is determined using the conventions cited above.

Normally a plot of $\ln(-\ln \theta)$ against $\ln t$ at various crystallization

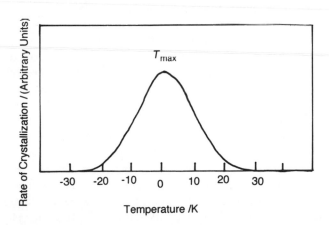

Figure 5.4. Schematic illustration of the rate of nucleation as a function of isothermal crystallization temperature

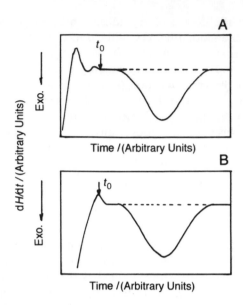

Figure 5.5. Conventions adopted to define the zero time (t_0) of isothermal crystallization. (A) The earliest point at which the sample baseline is flat. (B) The point at which the control light of a power compensation type DSC comes on

temperatures is used to calculate n (Figure 5.6). Frequently the error in estimating n is high, as the log–log plot is rarely linear. Even if the log–log plot is linear the details of the nucleation mechanism and growth geometry cannot be inferred solely by specifying the value of n. The degree of crystallinity of a system depends explicitly on the technique used to observe the system. Measurement of θ using X-ray diffractometry, DSC, NMR, FTIR, depolarization of transmitted light or changes in the sample density will result in a broad range of values, depending on the sample and the experimental conditions. Such variations in θ further undermine the significance of n. The isothermal crystallization of polymers can generally be described as a three-stage process. During the initial period no crystallization occurs and the length of this period depends on the crystallization temperature, the sample and the sensitivity of the detecting method. This is followed by an acceleration stage where widespread nucleation and growth occur. As neighbouring nuclei begin to impinge upon each other and the concentration of melt (or glassy material) diminishes, the rate of crystallization enters a pseudo-equilibrium stage which can continue for long periods without significantly increasing the crystallinity. The assumption that the transformation goes to completion is very rarely realized in polymer systems and the concept of the effective fraction of transformed material was introduced to alleviate this problem. However, the proportionality constant between the real and the effective fraction is arbitrarily assumed to be independent of time. Finally, polymer melts are not

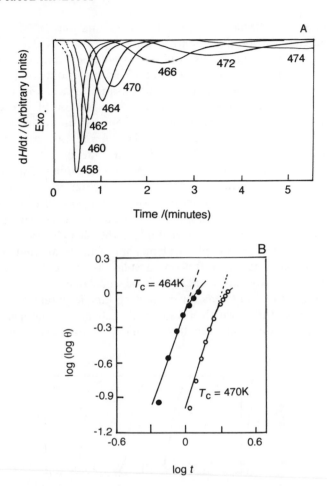

Figure 5.6. (A) DSC isothermal crystallization curves of carbon fibre-reinforced nylon 6 composite. At the lowest isothermal crystallization temperature nucleation occurs too rapidly to be accurately measured and must be inferred from the shape of the crystallization exotherm (dashed line). (B) The corresponding log–log plot, based on equation 5.26, from which the coefficient n can be calculated

isotropic solutions and the effect of the melt temperature on the isothermal crystallization behaviour of a sample should be established.

5.3.5 GENERAL COMMENT ON REACTION RATE KINETICS

The effect of experimental conditions on TG, DSC and DTA curves has been illustrated. Reaction kinetic parameters, whose estimates are based on the characteristics of these curves, are subject to the same influences. Standardization of the experimental conditions merely produces a uniform calculation error and does not alleviate the problem. Combinations of various experimental parameters can produce nearly identical TG curves. The product-like relationship between A and E observed in computer simulations based on the Arrhenius equation is presented in Figure 5.7 and is frequently ascribed to a compensation effect. A mathematical compensation effect, which is a trivial consequence of improper boundary conditions, is responsible. Quasi-isothermal, quasi-isobaric TG was specifically developed to minimize the effect of experimental conditions on the shape of TG curves, but even under these conditions the original authors[16] conceded that kinetic parameters estimated using the Arrhenius equation can have no physical meaning. The invalidity of this expression stems from the fact that the rate of reaction is not governed by the rate of chemical reaction, but by the considerably slower processes of heat and gas transport through the sample. Solid-state reactions generally depend on the environment immediately surrounding the particle and are therefore not activated. In addition, no physical meaning can be attributed to n in a solid-state reaction. Thermal diffusion in solids is an activated process for which in principle an activation energy can be calculated.

Among the proposed methods for estimating kinetic parameters from non-isothermal curves, those which presume the shape of the peak remains unchanged with heating rate are flawed and in most cases unusable. The

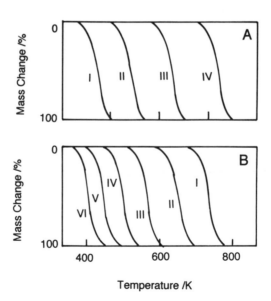

Figure 5.7. Calculated TG curves based on the Arrhenius relation. (A) For curves I–IV $E = 25$, 35, 45 and 55 kJ/mol, respectively. $A = 10^{14}$/s and $n = 1.05$. (B) For curves I–VI $A = 10^{10}$, 10^{12}, 10^{14}, 10^{16}, 10^{18}, and 10^{20}/s, respectively. $E = 40$ kJ/mol and $n = 1.05$. The so-called compensation effect is the result of improper boundary conditions

mathematical derivations which incorporate a temperature dependence for A are often incorrect as many of the series expansions used to integrate equation 5.18 are invalid for non-integer values of n. The Arrhenius equation was developed for isothermal reactions and in applying it to non-isothermal reactions invalid integration limits are sometimes chosen. If a linear plot is obtained using a particular method this does not signify that the assumptions of the derivation are satisfied, as many methods are insensitive to error and the shape of the TA curve is strongly influenced by experimental conditions. Also, a single TA peak does not mean that the reaction is simple as numerous parallel reactions may be occurring.

The effect of experimental conditions on the characteristics of the recorded data should be thoroughly investigated before any analysis is begun. Where the influence of experimental conditions is slight, reaction rate kinetic parameters can be estimated, but should not be assigned any physical significance and should only be used for comparison within a given set of samples. In all other cases the influence of experimental conditions on the sample behaviour can be used as a rich source of reliable information regarding the reaction processes. This information can be collated with data from other experimental techniques (e.g. EGA, thermomicroscopy) so that a truly representative picture of the physico-chemical processes occurring in the sample can be elucidated. The same is true for isothermal crystallization kinetics where investigation of the sample behaviour as a function of experimental conditions is frequently more rewarding than the estimation of meaningless coefficients.

5.4 Glass Transition of Polymers

The glass transition is exhibited by amorphous polymers or the amorphous regions of partially crystalline polymers when a viscous or rubbery state is transformed into a hard, brittle, glass-like state. The glass transition is neither a first- nor a second-order thermodynamic phase transition since neither the glassy state nor the viscous state is an equilibrium state. Relaxation phenomena are observed above and below the glass transition temperature (T_g). The glass transition of polymers is observed by DSC as a stepped increase in the heat capacity of the sample during heating due to an enhancement of molecular motion in the polymer. Measurement of T_g of polymers is an important practical application of TA. A market survey of Japanese companies in 1990 revealed that the primary reason for purchasing TA instruments was to determine T_g of manufacturing materials. The procedure to estimate T_g from a DSC curve is explained in Section 5.1.

Figure 5.8 shows DSC curves in the region of the glass transition of a series of atactic polystyrene films. Curve I reflects the thermal and mechanical history of a film prepared by melt pressing at 500 K and subsequently quenched to ambient temperature before being stored for some time. Enthalpy relaxation is revealed as a peak at the glass transition temperature. Curve II is

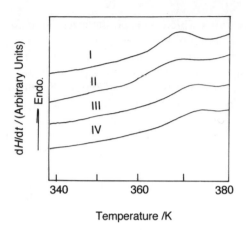

Figure 5.8. Effect of thermal history on the DSC heating curves of a series of atactic polystyrene films exhibiting a glass transition. The thermal histories are detailed in the text

the heating curve of a film cooled at a controlled rate from 400 to 310 K. The first and second scans of a film prepared by solvent casting (curves III and IV) are also presented. Generally T_g is measured from the second heating curve.

The glass transition can be observed in other amorphous polymers, such as poly(vinyl chloride), poly(vinyl acetate), poly(methyl acrylates), quenched poly(ethylene terephthalate) or quenched polycarbonates. A trace amount of organic solvent remaining in the sample following solvent casting shifts T_g to lower temperatures. Samples containing low molelcular mass compounds measured in open-type sample vessels may display a large endothermic peak immediately after the glass transition of the polymer. This peak is due to the vaporization of the low molecular mass component. If the sample baseline changes abruptly following the glass transition the value of T_g is not reliable. The sample should be dried until no further vaporization is observed and reheated to determine T_g. The recorded T_g depends on the heating rate because the glass transition is a relaxation phenomenon, and it is necessary to report the heating rate when presenting T_g data.

The glass transition may be difficult to measure for various reasons. T_g of a partially crystalline polymer, such as polyethylene or polypropylene, is diffi-cult to detect by DSC because enhancement of molecular motion in the amorphous regions is restricted by the crystalline regions. Restricted motion of the amorphous regions can be observed using other techniques such as DMA or NMR. In densely cross-linked polymers it is hard to observe the glass transition due to restriction of main-chain motion and the sample baseline step occurs over a broad temperature interval introducing a large error in the determination of T_g. Owing to widespread intramolecular and intermolecular hydrogen bonding it is difficult to measure T_g of many natural polymers in the dry state. In systems composed of several incompatible amorphous polymers it may be difficult to measure T_g of a minor component owing to its relatively low concentration. The phase diagrams of multi-component systems are frequently compiled from T_g data measured by DSC and the above difficulties may lead to inaccurate estimates of T_g.

5.4.1 ENTHALPY RELAXATION OF GLASSY POLYMERS

Polymers in the glassy state below the glass transition temperature are not in thermodynamic equilibrium and relax towards equilibrium with time. For this reason the experimentally measured enthalpy of glassy polymers decreases as a function of time if the sample is maintained below T_g. This phenomenon is called enthalpy relaxation and is monitored through the heat capacity change at the glass transition. In the presence of enthalpy relaxation, the mechanical, transport and other physical properties of the polymer vary as a function of temperature and time. Gas diffusion through polymer membranes can decrease by as much as two orders of magnitude owing to enthalpy relaxation at ambient temperature. The stress–strain curves of glassy polymers reveal more brittle behaviour as enthalpy relaxation proceeds.

The DSC curves of a quenched poly(thio-1,4-phenylenephenylphosphonyl-1,4-phenylenethio-4,4′-biphenylene) sample are presented in Figure 5.9. The quenched sample whose $T_{gi}{}'$ is 493 K was annealed at 482 K for various periods as indicated. The area of the endothermic peak increases with increasing annealing time.

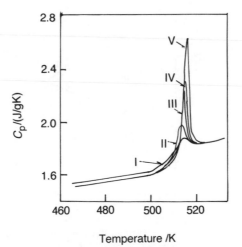

Figure 5.9. DSC heating curves of quenched poly(thio-1,4-phenylenephenylpho-sphonyl-1,4-phenylenethio-4,4′-biphenylene) after annealing at 482 K for the following periods: (I) 0; (II) 60; (III) 200; (IV) 305; (V) 1010 min

Precise analysis of enthalpy relaxation is not possible owing to the non-equilibrium nature of glassy polymers above and below the glass transition. Enthalpy relaxation can be characterized under certain limiting assumptions. If the viscous or rubbery state of the polymer above T_g is assumed to be an equilibrium state then the enthalpy of the supercooled state, formed by slow cooling, can be estimated by extrapolating the heat capacity in the viscous state to $T_g - 50$ K. This procedure is illustrated in Figure 5.10, from which the excess enthalpy, ΔH_0, can be calculated using

$$\Delta H_0 = \int_{T_a}^{T_g} C_{pv}(T)\,dT - \int_{T_a}^{T_g} C_{pg}(T)\,dT \qquad (5.27)$$

where C_{pv} and C_{pg} are the heat capacity in the viscous state and in the glassy state immediately after quenching, respectively. T_a is annealing temperature, $T_a = T_g - a$. In this example of polyoxyphenylene, $a = 15$ K. The enthalpy change for the annealed glassy polymer is given by

$$\Delta H_a = \int_{T_g-a}^{T_g+a} C_{pa}(T)\,dT - \int_{T_a-g}^{T_g+a} C_{pg}(T)\,dT \qquad (5.28)$$

where C_{pa} is the heat capacity of the annealed polymer. The total excess enthalpy, ΔH_t, is calculated from

$$\Delta H_t = \Delta H_0 - \Delta H_a \qquad (5.29)$$

ΔH_a increases with annealing time, decreasing ΔH_t and suggesting that the glassy state of the sample approaches equilibrium. The rate of change from the quenched state to the ideal equilibrium state can be characterized by a relaxation time, τ, expressed as

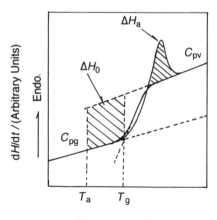

Figure 5.10. Enthalpy relaxation in glassy polymers is characterized by assuming that the enthalpy of the supercooled state can be determined from the extrapolated heat capacity in the viscous state, allowing the excess enthalpy to be estimated

$$\Delta H_t = -\Delta H_0 \exp(-t/\tau) \tag{5.30}$$

where t is the annealing time. Physically τ corresponds to a characteristic time for rearrangement of amorphous chains into more stable configurations. The C_p data in Figure 5.9 can be used to calculate ΔH_a and ΔH_t using equations 5.28 and 5.29. From the gradient of the plot of $\ln(\Delta H_t/\Delta H_0)$ against t, τ can be estimated (Figure 5.11).

ΔH_t can be estimated even if the gradients of C_{pv} and C_{pg} are not equal using the following equation instead of equation 5.27:

$$\Delta H_0 = a\Delta C_p \tag{5.31}$$

where ΔC_p is the C_p difference between the glassy state and the viscous state of the sample at T_g and is illustrated in Figure 5.12. In practice, T_g and ΔC_p are measured for the quenched sample. The sample is then annealed from $T_g - 10\,\mathrm{K}$ to $T_g - 20\,\mathrm{K}$ for various periods. Annealing can take a long time when $T_g - T_a$ is large. The DSC heating curves of the annealed samples are recorded under the same conditions as the quenched samples. The area of the curves from $T_g + a$ to $T_g - a$ of the quenched and the annealed samples are measured and the difference area calculated. The difference area corresponds to the ΔH_t. ΔH_a can be determined using equation 5.28, allowing ΔC_p and τ to be estimated. These procedures for characterizing enthalpy relaxation in glassy polymers are not thermodynamically rigorous, particularly the assumption that the viscous state above T_g is an equilibrium state. In some practical applications, including operational lifetime prediction, characterization of enthalpy relaxation by DSC has proved useful.

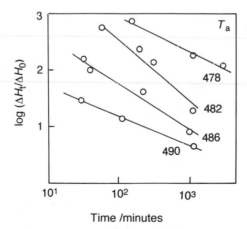

Figure 5.11. ln $\Delta H_t/\Delta H_0$ versus t at various annealing temperatures for poly(thio-1,4-phenylenephenylphosphonyl-1,4-phenylenethio-4,4′-biphenylene), from which a characteristic time for rearrangement of the polymer chains into more stable conformations (τ) can be calculated

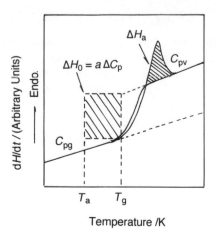

Figure 5.12. Enthalpy relaxation in glassy polymers can be characterized using the heat capacity difference between the glassy state and the viscous state at T_g

5.4.2 GLASS TRANSITION IN THE PRESENCE OF WATER

Polymers having hydrophilic components such as hydroxyl or amide groups form intermolecular bonds in the presence of water which strongly affect the characteristics of the glass transition. Main-chain motion is restricted owing to these intermolecular interactions and the glass transition temperature is higher than that of the hydrophilic polymer in the completely dry state or a similar hydrophobic polymer. In certain kinds of proteins and poly-saccharides no glass transition or melting is observed until decomposition of the main chain occurs because intramolecular and intermolecular hydrogen bonds stabilize the high-order structure of these polymers. On the other hand, introducing a small amount of water to a hydrophilic polymer may disrupt the intermolecular bonds, thereby enhancing the main-chain motion. In this case T_g shifts to lower temperatures in the presence of water. Hydro-philic polymers stored under ambient conditions contain a certain amount of bound water. In most practical applications the observed thermal and mechanical properties of the polymer reflect the presence of a nominal amount of water.

The relationship between the glass transition temperature and the water content of poly(4-hydroxystyrene) is summarized in Figure 5.13. The water content (W_c g/g) of the sample is defined as

$$W_c = \text{mass of water in sample/sample dry mass} \qquad (5.32)$$

The glass transition temperature of dry poly(4-hydroxystyrene) is 455 K and T_g decreases with increasing W_c. The value of T_g levels off around 370 K at a water content greater than 0.078 g/g. The levelling-off point agrees well with the bound water content calculated from the transition enthalpy of the water

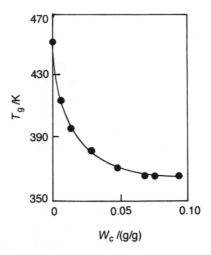

Figure 5.13. T_g as a function of water content for poly(4-hydroxystyrene)

in the sample (Section 5.11). The number of water molecules per hydroxyl group of poly(4-hydroxystyrene) can thus be estimated.

5.5 Heat Capacity Measurement by DSC

The differential heat supplied by a power compensation-type DSC instrument is proportional to the heat capacity of the sample, suggesting that C_p can be measured by DSC. The following details the steps of a C_p measurement using a power compensation-type instrument (Figure 5.14). (1) A pair of aluminium

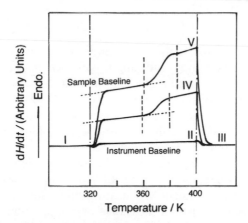

Figure 5.14. Heat capacity measurement using a power compensation-type DSC and sapphire as a standard reference material

sample vessels having very similar masses ($\Delta m \leqslant 0.01$ mg) are selected and one of them is placed in the sample holder of the DSC. (2) After powering-up, the DSC is maintained under a dry nitrogen gas flow for at least 60 min. The level of coolant in the reservoir is kept constant so that the instrument baseline is linear and very stable. (3) By maintaining the DSC system at an initial temperature (T_i) for 1 min, a straight line (curve I) is recorded. (4) Scanning at 5–10 K/min, the instrument baseline is measured (curve II). (5) By maintaining the DSC system at a final temperature (T_e) for 1 min, a straight line (curve III) is recorded. If the extrapolations of curves I and III are not co-linear, the slope control of the instrument is adjusted until this condition is satisfied and steps 3–5 are repeated. Once the above conditions have been satisfied, the slope, the horizontal and vertical axis sensitivities, the position of the zero point, the gas flow rate, the level of coolant, the orientation of the sample holder lid and the position of the recorder pen (if a chart recorder is being used) should be kept at those values for the duration of the experiment. (6) A 10–30 mg amount of a standard sapphire sample is weighed with a precision of ± 0.01 mg and placed in the second sample vessel (previously weighed). The sapphire sample is inserted into the sample holder and steps 3–5 are repeated to obtain curve IV. (7) The sapphire is removed from the sample vessel and replaced with the sample of known mass ± 0.01 mg. The sample is inserted into the sample holder and steps 3–5 are repeated to obtain curve V. The sample mass should be approximately 10 mg. The three curves II, IV and V should be coincident at T_i and T_e. If not, the measuring conditions for the sapphire and sample were not the same. For example, the gas flow rate may have changed during the experiment. The level of coolant in the reservoir must be kept constant. This condition is particularly difficult to satisfy if liquid nitrogen is used as a coolant. After correcting the difference in experimental conditions, the entire procedure is repeated.

C_p is calculated using the equation

$$C_{ps} = (I_s M_r / I_r M_s) C_{pr} \qquad (5.33)$$

where C_{ps} and C_{pr} are the sample and sapphire heat capacities, respectively, and M_s and M_r are the sample and sapphire masses, respectively. I_s and I_r are indicated in Figure 5.14. A computer can be used to measure I_r, I_s and C_{pr} at each sampling point and to calculate C_{ps}. Some software options have values of C_{pr} and I_r in memory and it is not necessary to measure curve VI. When the calculation is performed manually, I_r, I_s and C_{pr} are determined graphically at each temperature. After the calculation is completed, the C_p data from T_i to $T_i + 10$ K should be omitted since the stable heating condition is not attained at the initial stage of heating. The thermal history of the sample can be eliminated before the measurement by heating the sample to a temperature approximately 30 K greater than the transition temperature of the sample and maintaining that temperature for 5–10 min, while avoiding decomposition of the sample.

Figure 5.15 presents C_p data for atactic polystyrene. The original sample was quenched from 420 to 300 K and the other samples were annealed at 340 K for various times as indicated. By annealing at 340 K enthalpy relaxation is observed in these data.

Software options to measure C_p using quantitative DTA systems are available. A direct correlation between the difference temperature and C_p is assumed. Within the limits of this assumption reasonable data can be obtained. From a practical viewpoint the major difficulty is that the PID constants of the temperature programme must be altered in the course of the experiment to obtain the desired curves, which requires very good experimental technique on the part of the operator.

5.6 Purity Determination by DSC

The purity of a substance can be estimated by DSC using the effect of small amounts of impurity on the shape and temperature of the DSC melting endotherm. The procedure uses the van't Hoff equation:

$$\frac{1}{F_s} = \frac{\Delta H}{R} \frac{(T_0 - T_s)}{T_0^2} \frac{1}{X_2} \tag{5.34}$$

where T_s and T_0 (K) are the instantaneous sample temperature and the melting temperature of the pure substance, respectively, ΔH (J/mol) is the enthalpy of melting of the pure substance, X_2 is the mole fraction of impurity in the sample, R (J/mol K) denotes the gas constant and F_s is the fraction of sample melted at T_s and is given by

$$F_s = \frac{A_s}{A_T} \tag{5.35}$$

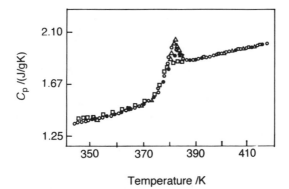

Figure 5.15. Heat capacity data for quenched atactic polystyrene. The samples were annealed at 340 K for the following periods before scanning: (□) 0; (●) 10; (○) 30; (△) 60 min

where A_T and A_s represent the total area of the endotherm and the area of the endotherm up to T_s, respectively. The validity of equation 5.34 is based on the following assumptions: (i) the melt is an ideal solution in which the impurities are soluble (eutectic system); (ii) melting occurs under conditions of thermodynamic equilibrium; (iii) the heat capacity of the melt is equal to that of the solid; (iv) in the solid state the impurities are not soluble in the principle component; (v) the principle component does not decompose or undergo any other polymorphic transitions at or near its melting temperature and the system is at constant pressure; (vi) there are no temperature gradients in the sample; (vii) the enthalpy of melting is independent of melting temperature; (viii) the impurity content is less than $5\,$mol-% so that the approximation $\ln X_1 \approx -X_2$ is true; (ix) $T_0^2 \approx T_s T_0$.

In practice, a small amount of sample (1–$3\,$mg) is heated (0.5–$1.25\,$K/min) in the DSC and the melting endotherm recorded. The endotherm is divided into segments whose onset temperature and area are measured. A plot of T_s against $1/F_s$ should, under ideal circumstances, yield a straight line whose intercept is T_0. From the slope of the line X_2 can be estimated using the equation

$$X_2 = \frac{\Delta H}{RT_0^2}\,(\text{slope}) \tag{5.36}$$

and the purity of the sample determined. However, the plot of T_s against $1/F_s$ is very often non-linear. Polymers are rarely (if ever) 100% crystalline and the presence of crystalline and amorphous regions means that the assumptions of the van't Hoff equation are not satisfied. In addition, the impurities in polymer systems are generally incorporated during polymerization and preparation, frequently forming solid solutions with the polymeric phase. Other parameters leading to non-linearity are thermal lag and undetected premelting of the sample. Some of the proposed solutions to these problems are discussed next.

5.6.1 THERMAL LAG

A DSC curve displays the differential heat supplied to the sample as a function of the programmed temperature while the difference between the programmed and measured sample temperatures is maintained below a predetermined value. Assuming ideal Newtonian behaviour of the DSC sample holder, the difference between the programmed temperature (T_p) and the true sample temperature (T_s) is given by

$$T_p - T_s = \frac{dH}{dt}\,R_0 \tag{5.37}$$

where R_0 is the thermal resistance of the DSC sample holder. By differentiating equation 5.37 with respect to time it can be shown that for a melting peak

$$\frac{d}{dT}\left(\frac{dH}{dt}\right) = \frac{1}{R_0} \tag{5.38}$$

From the slope of the melting endotherm of a pure material the thermal resistance of the sample holder can be determined (Figure 5.16). Using this value of R_0 the temperature scale of the sample DSC curve can be corrected. This correction slightly improves the linearity of the T_s against $1/F_s$ plot. R_0 should be calculated using a pure standard material whose melting temperature is as close as possible to that of the sample (Appendices 2.1 and 2.2).

5.6.2 UNDETECTED PREMELTING

Owing to the finite sensitivity of the DSC apparatus, premelting of the sample may go undetected, affecting the accuracy of the purity determination. The extent of premelting is difficult to quantify and a number of empirical solutions have been proposed to combat this problem. The fractional area can be rewritten in the form

$$\frac{1}{F_i} \approx \frac{A_T}{A_i + X} \qquad A_T \gg X \tag{5.39}$$

where X is an area added to the segment area so that the plot of T_s against $1/F_s$ becomes linear. The boundary conditions are that $(A_T + X)$ can be no greater than ΔH and that the intercept on the vertical axis correspond to T_0, if ΔH and T_0 are known. Sometimes X is a large fraction of A_T and in this case equation 5.39 is not appropriate. An alternative approach[17] uses the fact that the coordinates of a point on the plot of T_s against $1/F_s$ are $(A_T/A_i, T_i)$ and a value of X is required so that all points lie on the same straight line with coordinates $[(A_T + X)/(A_i + X), T_i]$. For any three points on the line

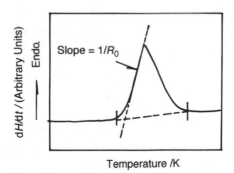

Figure 5.16. Thermal resistance of sample holder estimated from the melting endotherm of a pure compound

$$\frac{T_3 - T_2}{\dfrac{A_T + X}{A_3 + X} - \dfrac{A_T + X}{A_2 + X}} = \frac{T_2 - T_1}{\dfrac{A_T + X}{A_2 + X} - \dfrac{A_T + X}{A_1 + X}} \tag{5.40}$$

and rearranging

$$X = \frac{\dfrac{(T_3 - T_2)A_3}{T_2 - T_1} - \dfrac{(A_3 - A_2)A_1}{A_2 - A_1}}{\dfrac{A_3 - A_1}{A_2 - A_1} - \dfrac{T_3 - T_2}{T_2 - T_1}} \tag{5.41}$$

The three points should be chosen from the extremities and middle of the curve and with the improved linearity T_0 and X_2 can be estimated. This method can be extended and applied to all points on the T_s against $1/F_s$ plot. The boundary conditions are the same as those of equation 5.39.

5.6.3 GENERAL COMMENT ON PURITY DETERMINATION BY DSC

The ideal behaviour assumed in deriving the Van't Hoff equation is generally not observed and the measured impurity concentration is strongly dependent on the nature of the impurity. The effect of low boiling point solvent impurities such as water may not be detected if they vaporize before melting occurs. A DSC purity measurement is not performed under equilibrium conditions and is therefore only approximate. The estimate should be verified by comparison with values from other techniques such as high-performance liquid chromatography (HPLC). For purity measurements the energy and temperature calibration of the DSC system should be as precise as possible. Allowance must be made for the difference between the instrument and sample baselines when estimating A_s and T_s (Figure 5.17).

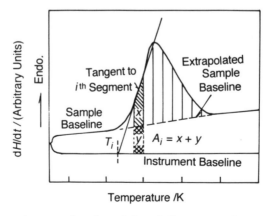

Figure 5.17. Correction to estimation of A_s and T_s necessary because of the difference between the sample baseline and the instrument baseline

5.7 Crystallinity Determination by DSC

The measured crystallinity of a polymer has no absolute value and is critically dependent on the experimental technique used to determine it. An estimate of the crystallinity of a polymer can be made from DSC data assuming strict two-state behaviour. In this case the polymer is presumed to be composed of distinct, non-interacting amorphous and crystalline regions where reordering of the polymer structure only occurs at the melting temperature of the crystalline component. Despite the obvious limitations of this model, it is widely used in industry to determine the crystallinity of polymers. The crystallinity (X_c) is calculated using

$$X_c = \Delta H/\Delta H_{100} \qquad (5.42)$$

where ΔH and ΔH_{100} are the measured enthalpy of melting of the sample and the enthalpy of melting of a 100% pure crystalline sample of the same polymer, respectively. For most polymers, samples whose crystallinity is even approximately 100% are not available and ΔH_{100} is replaced by the enthalpy of fusion per mole of chemical repeating units (ΔH_u). ΔH_u is calculated using Flory's relation[18] for the depression of the equilibrium melting temperature (T_m^0) of a homopolymer due to the presence of a low molecular mass diluent:

$$1/T_m - 1/T_m^0 = (R/\Delta H_u)(V_u/V_1)(v_1 - x_1v_1^2) \qquad (5.43)$$

where T_m is the melting temperature of the polymer–diluent system, V_u and V_1 are the molar volumes of the repeating unit and the diluent, respectively, v_1 is the volume fraction of the diluent and x_1 is the thermodynamic interaction parameter. Values of ΔH_u for some polymers are available[19]. Where ΔH_u is unknown an alternative method for determining ΔH_{100} must be found.

Figure 5.18 presents the calculated crystallinity of poly(ethylene terephthalate) as a function of annealing temperature using DSC and X-ray and IR spectroscopy data. It can seen that the estimates of X_c vary greatly. DSC is clearly the least sensitive to the effect of annealing on the sample crystallinity. This is because reordering of the polymer structure occurs during the DSC measurement.

5.8 Molecular Rearrangement During Scanning

The high-order structure of polymers can undergo many kinds of transformation during scanning. Figure 5.19 presents DSC curves of poly(ethylene terephtalate) (PET). Curve I shows the sample heated at 10 K/min where a melting peak is observed at 529 K. The sample is subsequently cooled at 10 K/min and a crystallization exotherm is recorded at 468 K (curve II). By reheating at the same rate a sub-melting peak is revealed at a temperature

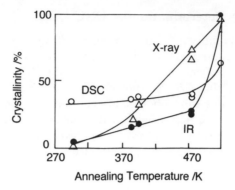

Figure 5.18. Crystallinity of poly(ethylene terephthalate) as a function of annealing temperature determined using X-ray diffractometry, IR spectroscopy and DSC

lower than main melting peak (curve III). The area of the sub-melting peak increases with increasing heating rate, suggesting that the crystalline regions of PET are reorganized during scanning. The crystallites formed during rapid heating melt at lower temperatures, indicating that defects and irregular molecular arrangements are present. When quenched from the molten state to 273 K, PET freezes in a glassy state and an amorphous halo pattern is observed by X-ray diffractometry.

Curve IV is the heating curve of quenched PET. A glass transition, cold crystallization, premelt crystallization and melting are observed. The polymer chains attain sufficient mobility following the glass transition to begin the

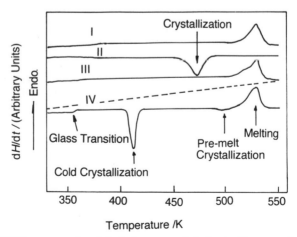

Figure 5.19. DSC curves of poly(ethylene terephthalate): (I) as received sample, (II) cooled at 10 K/min (III) heated at 10 K/min, and (IV) heated at 10 K/min following quenching

formation of crystallites in the region of the cold crystallization exotherm. The enthalpy change involved during rearrangement is responsible for the DSC peak and the crystallinity of PET determined by X-ray analysis is low at this temperature. With continued heating the crystallites are annealed and the crystallinity increases so that a melting peak is observed. X-ray diffraction data reveal that crystallization is enhanced in the premelt crystallization temperature region.

5.9 Polymorphism

Polymorphism is the term used to describe the occurrence of different structural forms of a material, and is observed in polymers such as polyamides, polypropylene, polysaccharides and fluorinated polymers. X-ray diffractometry is the principal technique used to probe the polymorphic nature of polymers while the temperature and enthalpy change associated with crystal to crystal transitions are measured using DSC.

Polypropylene (PP) has two crystalline forms. A monoclinic crystal (α-type) is obtained by slow crystallization from the molten state and a hexagonal crystal (β-type) is formed by annealing while a temperature gradient is maintained across the sample. Figure 5.20A presents DSC heating curves of an isotactic PP film prepared by temperature gradient annealing[20]. The film was pressed at 473 K and cooled to 443 K, where the pressure was released. The film was then annealed by sandwiching it between metal plates, one of which was maintained at 438 K and the other at 338 K. The β-type transcrystal was obtained where the **a** molecular axis corresponds to the direction of the temperature gradient. By heating at the fastest rate two melting endotherms are observed at 420 and 430 K. The DSC curve recorded at 20 K/min reveals an additional peak at 439 K and the area of this endotherm increases with decreasing heating rate. An endotherm and an exotherm are observed between 420 and 430 K in the curve measured at 1.25 K/min. The melting peak at 420 K is asymmetric whereas the exotherm is sharp. The endotherm at 420 K is attributed to the melting of β-type crystals. At a low heating rate recrystallization begins during melting of the β-form, producing an exothermic peak. The endotherm due to melting of the β-form and the exotherm due to the recrystallization occur simultaneously. The deconvolution of the DSC curve recorded at 10 K/min is shown schematically in Figure 5.20B. The temperature and enthalpy of the β-crystal melting peak and the recrystallization peak cannot be reliably determined under these conditions. The endotherm observed in the region of 424 K is the continuation of the melting peak at 420 K. The melting peaks at 430 K and at 439 K are attributed to melting of α-type crystals and melting of recrystallized α-type crystals, respectively. At high heating rates only the melting peaks of β-type and α-type crystals can be observed because there is insufficient time for recrystallization to occur.

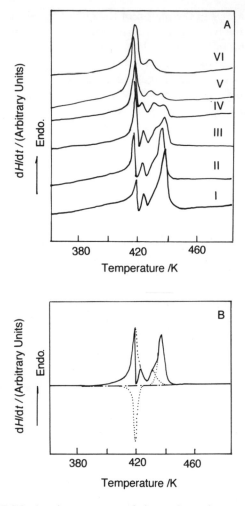

Figure 5.20. (A) DSC heating curves of isotactic polypropylene, prepared by temperature gradient annealing, as a function of scanning rate: (I) 1.25; (II) 2.5; (III) 5; (IV) 10; (V) 20; (VI) 40 K/min. (B) Deconvolution of the DSC curve recorded at 10 K/min. (Reproduced by permission of the Japan Society of Calorimetry and Thermal Analysis from *Netsu Sokutei*, **16**, 58, (1989))

5.10 Annealing

Isotactic polystyrene (iso-PSt) forms a glassy state when it is quenched from the molten state to 200 K. The DSC heating curve of quenched iso-PSt reveals a glass transition, cold crystallization and melting. By annealing at various temperatures a sub-melting peak can be observed. Figure 5.21A shows DSC melting curves of annealed iso-PSt. The temperature of the sub-peak increases with annealing temperature. The DSC melting curve of iso-PSt following

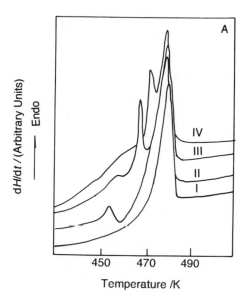

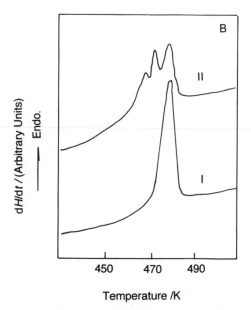

Figure 5.21. DSC heating curves of isotactic polystyrene. (A) The sample was annealed for 10 min at (I) 403, (II) 423, (III) 453 and (IV) 463 K. (B) (I) Annealed at 403 K for 10 min; (II) multi-step annealing. The sample was annealed for 2 min at 463 K and quenched to 443 K, where it was annealed for 2 min before being quenched to 310 K

multi-step annealing is presented in Figure 5.21B. Two sub-peaks are observed corresponding to each annealing step.

5.11 Bound Water Content

Owing to the effect of water on the performance of commercial polymers and the crucial role played by water–polymer interactions in biological processes, hydrated polymer systems are widely investigated. In the presence of excess water a polymer may become swollen, exhibiting large changes in mechanical and chemical properties. Water can plasticize the polymer matrix or form stable bridges through hydrogen bonding, resulting in an anti-plasticizing effect. The behaviour of water can be transformed in the presence of a polymer depending on the degree of chemical or physical association between the water and polymer phases.

Water whose melting/crystallization temperature and enthalpy of melting/crystallization are not significantly different from those of normal (bulk) water is called freezing water. Those water species exhibiting large differences in transition enthalpies and temperatures, or those for which no phase transition can be observed calorimetrically, are referred to as bound water. It is frequently impossible to observe crystallization exotherms or melting endotherms for water fractions very closely associated with the polymer matrix. These water species are called non-freezable. Less closely associated water species do exhibit melting/crystallization peaks, but often considerable super-cooling is observed and the area of the peaks on both the heating and cooling cycles are significantly smaller than those of bulk water. These water fractions are referred to as freezing-bound water. The sum of the freezing-bound and non-freezing water fractions is the bound water content.

These different water species are illustrated in Figure 5.22, which presents the DSC crystallization curves for water sorbed on poly(4-hydroxystyrene). The water content is given by the mass of water in the polymer divided by the dry mass of the sample, expressed in units of g/g. At the lowest water content no exothermic peak is observed. All of the water in the polymer at this water concentration is non-freezing water. At a higher water content a freezing exotherm is observed at 225 K whose area is considerable smaller than 333 J/g, which is the enthalpy of crystallization of bulk water. This peak is due to freezing-bound water in the system. At the highest water content a large exotherm is observed in the region of 273 K whose enthalpy of transition is close to that of bulk water. This exotherm is ascribed to the crystallization of the freezing water in the hydrated polymer.

The total water content of the system is given by

$$W_T = W_f + W_{fb} + W_{nf} \tag{5.44}$$

where W_f, W_{fb} and W_{nf} are the freezing, the freezing-bound and the non-freezing water contents, respectively. Determination of the exact proportions of

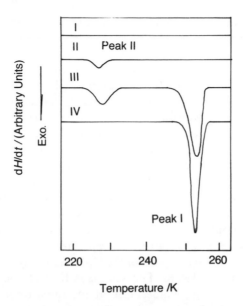

Figure 5.22. DSC crystallization curves of water sorbed on poly(4-hydroxystyrene). (I) $W_T = 0.079$ g/g; (II) $W_T = 0.107$ g/g; (III) $W_T = 0.263$ g/g; (IV) pure water

these water species in a hydrated polymer is an important step in understanding the physico-chemical processes which govern the behaviour of the system.

5.11.1 EXPERIMENTAL PROCEDURE

DSC is commonly used to determine the proportions of the various water fractions present in hydrated polymers. Sample vessels which can be sealed hermetically are required. If aluminium sample vessels are to be used they should first be placed in an autoclave with a smalll amount of pure water at 373 K for 3–5 h to eliminate the formation of aluminium hydroxide on the inner surfaces of the sample vessel during the measuring cycle. Almost all polymers contain a small amount of water which is absorbed during synthesis, processing or storage. When closely associated with the polymer matrix this water can remain in the matrix even after heating the polymer to 373 K under reduced pressure. It is important to establish the concentration of this water species so that the total amount of water present in the sample after hydration is precisely known. To determine this intrinsic water content the sample should be weighed as accurately as possible, noting that the sample will absorb water from the atmosphere during weighing. A microbalance with sensitivity $\geqslant 0.001$ mg is necessary. The sample vessel is pierced, quickly placed in the DSC at room temperature and heated at 10 K/min. An endothermic deviation in the sample baseline due to the vaporization of water is observed. The heat of vaporization of water is high (2257 J/g) and the presence of very small amounts of water can be detected by this procedure. The sample is heated until no deviation in the sample baseline is observed. The dried sample is then quickly reweighed and the intrinsic water content determined.

The following procedure is recommended to obtain a uniformly hydrated sample. A precisely known amount of sample is placed in a sample vessel and an excess amount of distilled, deionized water is added to the sample using a microsyringe. While monitoring the total mass of the sample vessel and hydrated polymer, the excess water is allowed to evaporate until the desired water concentration is achieved. The sample vessel is then hermetically sealed and allowed to equilibrate for 1–7 days. The storage temperature should be greater than the glass transition temperature of the dry polymer and is normally in the range 280–365 K. Natural polymers are prone to acid hydrolysis, resulting in a reduction of molecular mass and equilibration of hydrated natural polymers should be carried out at temperatures $\leqslant 285$ K. The equilibration period is longest for hydrophobic polymers.

The equilibrated sample is placed in the DSC at the storage temperature and cooled at 5–10 K/min to 150 K. The sample is held at 150 K for 15 min and heated back to the storage temperature at the same rate. This procedure is repeated three times, the heating and cooling thermograms being recorded each time. The temperature and number of crystallization exotherms observed depend on the nature of the polymer and the water concentration. From the cooling cycle data the proportion of freezing water, W_f, is calculated by dividing the total area of the freezing water peak (peak I in Figure 5.22) by the heat of crystallization of bulk water. The reported value should be the average of the estimates from the three thermal cycles. The heat of crystallization is not constant for all water species and therefore W_{fb} cannot be determined in the same way. Instead, the total area of the freezing-bound water peak (peak II in Figure 5.22) per gram of dry polymer should be plotted as a function of water content. The intercept of the linear plot is the amount of non-freezing water in the hydrated polymer, W_{nf}, and the slope is the enthalpy of crystallization of the freezing-bound water which can be used to calculate W_{fb}.

It is not recommended to use the heating cycle data to measure the bound water content as the area of the endothermic peak does not represent the enthalpy change associated with the transition from ice to water, but rather the change in enthalpy associated with the transformation from water in a crystalline state to a homogeneous mixture of water and polymer. The difference is the heat of mixing of water with the polymer, which is very difficult to estimate. In addition, owing to the complex interplay between the ice structures present, the non-freezable water fractions and the mobile elements of the polymer matrix, clear resolution of the different water species during heating is often impossible.

In some cases it is possible to measure the bound water content of a hydrated polymer using TG. The loss of freezable water occurs from room temperature onwards and a relatively large amount of water evaporates during handling.

These losses, coupled with the losses which occur during the preliminary heating cycle in the isothermal mode, render estimates of the total water content by TG less reliable than those from DSC. The bound water fractions are less prone to evaporation during handling and can be determined from the TG curve. With the DTG curve it is sometimes possible to resolve the peaks due to the non-freezing and freezing-bound water and to estimate W_{nf} and W_{fb}.

5.12 Phase Diagram

A phase diagram is a graphical representation of the relationship between a given set of experimental parameters and the phase changes occurring in a material. Sample volume, transition temperature and enthalpy, pressure and composition of the material are commonly used parameters in phase diagrams. Transition temperatures measured by TA are not equilibrium values and vary with the experimental conditions, particularly the scanning rate. Therefore, when presenting a phase diagram compiled from TA data the experimental conditions must be described in detail.

Xanthan gum is an anionic polysaccharide secreted by certain bacteria which in the dry state does not exhibit a first-order phase transition. In the presence of a small amount of water a glass transition, cold crystallizaton, melting and a liquid crystal transition are observed. Figure 5.23A presents DSC heating curves of water–xanthan gum systems with various water contents. A 3 mg sample was hermetically sealed in an aluminium sample vessel, cooled from 320 to 150 K at 10 K/min and subsequently heated at 10 K/min. With reference to Section 5.1, the transition temperatures are defined as follows: glass transition temperature T_{ig} and melting and crystallization temperatures T_{pm} and T_{pc}, respectively. The corresponding phase diagram, showing the transition temperatures as a function of the water content (W_c) for the water–xanthan gum system, is presented in Figure 5.23B. The melting temperature increases with W_c levelling off at $W_c = 1.4$ g/g. The glass transition temperature decreases in the W_c range where freezable water (Section 5.11) is no longer present. The liquid crystal transition is observed between $W_c = 0.45$ and 1.0 g/g in the temperature range 260–300 K. The liquid crystalline nature of water–xanthan gum systems can also be observed under the same conditions by thermomicroscopy.

5.13 Gel–Sol Transition

Polymer chains can form infinite networks by either physical or chemical association. Reversible networks in the presence of a solvent form reversible gels. The cross-links between individual chains in a reversible gel are localized, but not permanent, and the interacting groups dissociate and reassociate according to the conditions of thermodynamic equilibrium. The gel structure present at low temperatures is transformed on heating and a liquid state is

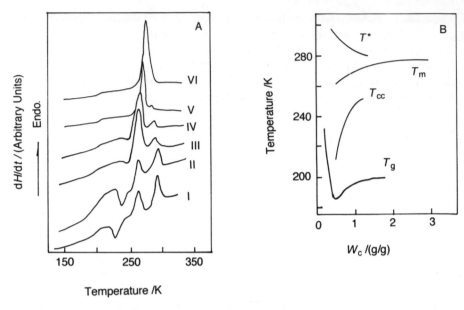

Figure 5.23. (A) DSC heating curves for water–xanthan gum systems at various water concentrations: (I) 0.54; (II); 0.57; (III) 0.70; (IV) 0.84; (V) 1.06; (VI) 1.40 g/g. (B) Phase diagram compiled from DSC heating curves. T_g, Glass transition; T_{cc}, cold crystallization; T_m, melting; T^*, transition from mesophase to liquid state

observed. This process, which can be reversed on cooling, is called the gel–sol transition and can be monitored by HS-DSC. The HS-DSC curve of the gel–sol transition is often very broad and structured on both heating and cooling. Hysteresis is generally observed between the heating and cooling cycles. For many gels there is no strict gel point, rather the gel–sol transition represents a progressive transformation from an elastic state (gel) to a viscous state (sol). The gel–sol transition is influenced by the molecular mass and polydispersity of the polymer and also by the nature, concentration, ionic content and pH of the solvent. The presence of small amounts of impurities can also affect the transition characteristics.

When measuring the enthalpy of transition by HS-DSC, it is important to establish the most appropriate extrapolated sample baseline. It is generally assumed that the sample baselines observed before and after the transition can be expressed as linear functions of temperature, and can be extrapolated into the transition region. For a system which is strictly two-state (A ⇌ B), the apparent specific heats are given by

$$C_A = X + YT \tag{5.45}$$

$$C_B = W + ZT \tag{5.46}$$

where W, X, Y and Z are constants and T is the temperature. The extrapolated sample baseline in the region of the transition is a curve which changes from one apparent specific heat to the other as a function of the degree of conversion and is given by

$$C_{ESB} = (1 - \alpha)(X + YT) + \alpha(W + ZT) \tag{5.47}$$

where α is the degree of conversion and the constants are determined graphically from the HS-DSC curve. Figure 5.24 shows an extrapolated sample baseline calculated using this method. The enthalpy of transition can then be estimated using

$$\Delta H = Q + (W - X)T + 0.5(Z - Y)T^2 \tag{5.48}$$

where

$$Q = \Delta H_{\frac{1}{2}} - (W - X)T_{\frac{1}{2}} + (Z - Y)T_{\frac{1}{2}}^2 \tag{5.49}$$

$T_{\frac{1}{2}}$ and $\Delta H_{\frac{1}{2}}$ are the temperature of half conversion and the enthalpy of transition at half conversion, respectively.

The ideal behaviour assumed in calculating the enthalpy of transition is rarely observed in gel systems. Neither the gel nor the sol state are equilibrium states and therefore $d(\Delta H)/dT$ cannot be directly correlated with ΔC_p for the transition. The sol state is not an isotropic liquid state, particularly in the case of DNA and polysaccharides, as high-order structures can be formed in the sol state, greatly affecting the gel structure subsequently formed on cooling. In the case where the polymer behaves as a linear polyelectrolyte, there is a contribution to the enthalpy of transition from the difference in the linear charge density arising from the change in conformation of the mole-

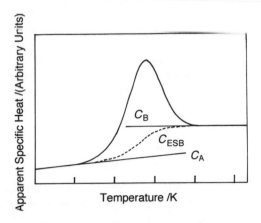

Figure 5.24. Calculated extrapolated sample baseline for HS-DSC heating curve, using equation 5.47

cule. The extrapolated sample baseline calculated in the above manner is therefore approximate.

The shape of the HS-DSC curve is often analysed in support of a particular theory of gelation. These procedures should be used with caution. First, the gel–sol transition is recorded under non-equilibrium conditions irrespective of the heating (cooling) rate. Software corrections are frequently applied to the HS-DSC curves to improve the linearity of the sample baseline, thereby affecting the peak shape. Gelation models frequently assume strict two-state behaviour of the system neglecting the rigidity of the junction zones (cross-link areas in physical gels which maintain the integrity of the gel) and ignoring all network imperfections. Under these conditions any agreement between the shapes of the theoretical and observed curves is fortuitous. For example, assuming strict two-state behaviour of the polysaccharide schizophyllan, each triple helix in the gel state transforms into three random coils in the sol state. The correlation between the theoretical and observed HS-DSC curves is very good, but the calculated Van't Hoff enthalpy (Section 5.13.1) is approximately three times too large. The discrepancy arises because the assumptions inherent in strict two-state behaviour are only applicable to short oligomers and not to polymers. In this case the transition is controlled by the denaturation of the individual helices and numerous intermediate states are formed. It is worth noting that not all HS-DSC instruments are the same and that the shape of a difference HS-DSC curve is different from a derivative HS-DSC curve. Where the gel—sol transition is more complicated than a simple two-state transition analysis of the shape of the HS-DSC curve is not recommended.

The gel–sol transition can also be monitored by mechanical analysis either by measuring the shear modulus as a function of temperature or by compiling a mastercurve from isothermal viscoelastic measurements over a range of frequencies.

5.13.1 OTHER APPLICATIONS OF HS-DSC

HS-DSC can also be used to study the denaturation of proteins, protein folding, helix—helix transitions and the motion of side-chains. A comparison of the transition with strict two-state behaviour can be made by comparing the observed HS-DSC curve with the theoretical curve derived using the van't Hoff relationship

$$\frac{\delta \ln K}{\delta T} = \frac{\Delta H_{\mathrm{vH}}}{RT^2} \tag{5.50}$$

where K is the equilibrium constant of the ideal two-state reaction and ΔH_{vH} is the van't Hoff enthalpy. K is determined from the degree of conversion, α:

$$K = \frac{\alpha}{(1 - \alpha)} \tag{5.51}$$

The van't Hoff enthalpy is calculated from

$$\Delta H_{vH} = 4RT_{\frac{1}{2}}^{2}(C_{\frac{1}{2}}/\Delta H) \tag{5.52}$$

where $T_{\frac{1}{2}}$ and $C_{\frac{1}{2}}$ are the temperature of half conversion and excess specific heat at half conversion, respectively, R is the gas constant and ΔH the enthalpy of transition given by equation 5.48. The excess specific heat is estimated using

$$C_{EX} = \Delta H \frac{d\alpha}{dT} \alpha(1 - \alpha)\beta \frac{(\Delta H)^2}{RT^2} \tag{5.53}$$

$\beta = \Delta H_{vH}/\Delta H$ and is assumed to be independent of temperature. If a plot of β/M_w against the appropriate experimental parameter (for example, pH of solution, water concentration) is very close to 1.00, the transition is considered to be two-state. A value greater than 1.00 suggests that intermolecular interactions are occurring and less than 1.00 indicates that an intermediate state(s) is formed during the transition.

5.14 References

[1]   Hatakeyama, T. and Kanetsuna, H. *Thermochimica Acta* **138**, 327 (1989).
[2]   Takahashi, Y. *Thermochimica Acta* **88**, 199 (1985).
[3]   Freeman, E.S. and Carroll, B. *Journal of Physical Chemistry* **62**, 394 (1958).
[4]   Jerez, A. *Journal of Thermal Analysis* **26**, 315 (1983).
[5]   Van Dooren, A.A. and Müller, B.W. *Thermochimica Acta* **65**, 269 (1983).
[6]   Kissinger, H.E. *Analytical Chemistry* **29**, 1702 (1957).
[7]   Augis, J.A. and Bennett, J.E. *Journal of Thermal Analysis* **13**, 283 (1978).
[8]   Elder, J.P. *Journal of Thermal Analysis* **30**, 657 (1985).
[9]   Borchardt, H.J. and Daniels, F. *Journal of the American Chemical Society* **79**, 41 (1957).
[10] Eyraud, C. *Comptes Rendus de Recherches* **238**, 1511 (1954).
[11] Doyle, C.D. *Journal of Applied Polymer Science* **5**, 285 (1961).
[12] Ozawa, T. *Bulletin of the Chemical Society of Japan* **38**, 1881 (1965).
[13] Kassman, A.J. *Thermochimica Acta* **84**, 89 (1985).
[14] Coats, A.W. and Redfern, J.P. *Nature (London)* **201**, 68 (1964).
[15] Flynn, J.H. and Dickens, B. *Thermochimica Acta* **15**, 1 (1976).
[16] Arnold, M., Veress, G.E., Paulik, J. and Paulik, F. *Journal of Thermal Analysis* **17**, 507 (1979); and *Analytica Chimica Acta* **124**, 341 (1981).
[17] Sondack, D.L. *Analytical Chemistry* **44**, 888 (1972).
[18] Flory, P.J. *Principles of Polymer Chemistry*, Cornell University Press, Ithaca, NY (1953).
[19] Mandelkern, L. *Crystallization of Polymers*, McGraw-Hill, New York (1964).
[20] Yamamoto, Y., Nakazato, M. and Saito, Y. *Netsu Sokutei* **16**, 58 (1989).

6 OTHER THERMAL ANALYSIS METHODS

6.1 Evolved Gas Analysis

Evolved gas analysis (EGA) is the general term for any technique which determines the nature and amount of volatile products evolved by a sample as it is subjected to a controlled temperature programme. EGA was preceeded by evolved gas detection (EGD), which merely detected the presence of evolved gases. When used in tandem with TG or DTA, EGA is primarily employed to determine the composition and concentration of evolved gases from mass loss reactions. Parallel and overlapping reactions which often result in a single feature on a TA curve can be resolved by identifying the associated volatile product, and in some cases quantitative information about the decomposition reaction rate can be obtained. Evolved gases can be sampled either continuously or intermittently. The two most common methods of EGA, mass spectroscopy (MS) and Fourier transform infrared (FTIR) spectroscopy, continuously monitor the purge gas as a function of time or temperature. Gas chromatography (GC) is an example of an intermittent sampling technique, where a fraction of the purge gas is collected over a given time or temperature interval and subsequently analysed.

In a coupled TA–EGA configuration the evolved gases should be analysed as quickly as possible after release from the sample to avoid secondary gas-phase reactions and condensation. This is particularly important when there is a large temperature difference between the sample and the gas analyser. The connecting stage between the instruments should be inert. Diffusion broadening, due to the increased volume of the combined system, can reduce the spectral resolution of the evolved gas. Selection of the appropriate purge gas and flow rate are important. Owing to its low mass, high thermal conductivity and chemical inertness, helium is commonly employed as the purge gas for coupled TA–EGA systems. Other common purge gases include argon and hydrogen. The selectivity of the analyser should also be considered. For example, FTIR does not detect non-polar molecules (H_2, N_2, O_2).

6.1.1 MASS SPECTROMETRY (MS)

MS is a high-sensitivity, non-specific technique used to identify unknown compounds. When bombarded by electrons all substances ionize and fragment in a unique manner. The mass spectrum, which records the mass and relative abundance of the ion fragments, gives a fingerprint for each compound. MS, using quadrupole mass spectrometers, is the most commonly used EGA technique. A TG–MS instrument is presented in Figure 6.1. The evolved gas components are detected with almost equal sensitivity provided they remain in the gaseous state at the temperature and pressure in the vicinity of the ion source. The entire mass spectrum, or selected regions of the spectrum, can be monitored continuously and the amount of sample can be of the order of nanograms. The greatest difficulty in coupling a mass spectrometer with a TA instrument is the very large pressure difference between the instruments. A range of coupling valves are available so that only a small fraction of the purge gas enters the ion source, allowing the high vacuum of the mass spectrometer to be maintained. Figure 6.2 shows the decomposition of poly(ethylene–co-vinyl alcohol) as studied using simultaneous TG MS.

6.1.2 FOURIER TRANSFORM INFRARED (FTIR) SPECTROSCOPY

When IR radiation $(0.7 < \lambda < 500 \, \mu m)$ impinges upon a molecule, the absorption pattern in certain frequency regions can be correlated with specific stretching and bending motions in the molecule. Thus, by examination of the IR absorption spectrum it is possible to identify the molecular species. Although more selective than MS, FTIR is widely employed in EGA, owing to its relatively high sensitivity and short spectrum acquisition time. The structure

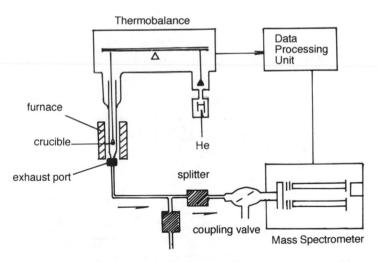

Figure 6.1. Schematic diagram of a simultaneous TG–MS apparatus. The separator ensures that the high vacuum of the quadrupole mass spectrometer is maintained. Owing to the high sensitivity of the mass spectrometer only a small fraction of the evolved gas is analysed (courtesy of Seiko Instruments)

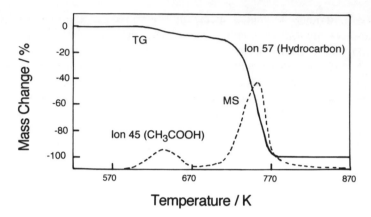

Figure 6.2. Decomposition of poly(ethylene–co-vinyl alcohol) as monitored using TG-MS (courtesy of Seiko Instruments)

of a TG–DTA–FTIR instrument is shown in Figure 6.3. For optimum performance the lowest purge gas flow rate possible is recommended to increase the concentration of product gases, while avoiding secondary gas-phase reactions. Corrosive and reactive decomposition products are more easily handled by the TG–FTIR coupling mechanism than by TG–MS. In Figure 6.4 the decomposition of poly(ethylene terephthalate) as revealed using TG–DTA–FTIR is shown.

6.1.3 GAS CHROMATOGRAPHY (GC)

In GC volatile products, carried by a purge gas, are absorbed at the head of the chromatographic column by the column material, and subsequently

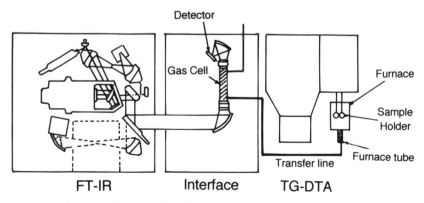

Figure 6.3. Schematic diagram of an integrated TG–DTA–FTIR apparatus (courtesy of Seiko Instruments)

desorbed by fresh purge gas. This sorption–desorption process occurs repeatedly as the volatile products are swept through the column. Each component passes through the column at a characteristic rate and the components are eluted in order of increasing partition ratio. At the column outlet the compo-

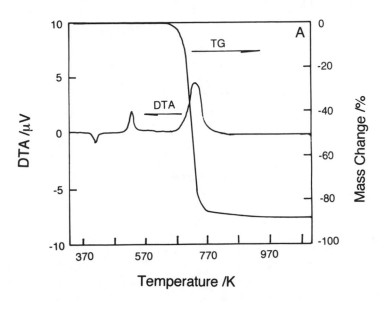

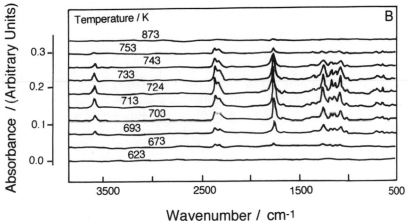

Figure 6.4. Decomposition of poly(ethylene terephthalate) as recorded using TG–DTA–FTIR. (A) Simultaneous TG–DTA curves. (B) IR absorption spectra of the evolved gases at various temperatures. (C) Specific gas profiles of the evolved gases. The integrated IR absorption spectra are plotted as a function of temperature. Each wavelength interval monitors the evolution of a particular compound. (I) Benzoic acid; (II) Carbon dioxide; (III) Aromatic carboxylic acid; (IV) Aromatic esters; (V) Carbon monoxide

(continued)

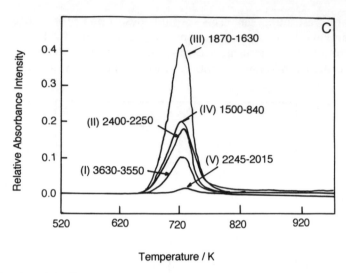

Figure 6.4. *(continued)*

sition of the purge gas is determined as a function of time, usually by measuring the thermal conductivity or by flame ionization analysis. The emergence time of a GC peak is unique to each component and the peak area is proportional to the concentration of that component. GC can only be used intermittently because several minutes are required for the components with the longest retention times to leave the column. With a suitable choice of column material the gas components can be separated and identified, although repeated samplings are sometimes necessary for unequivocal assignment. Selection of the appropriate column temperature is important to avoid poor resolution of low boiling point components. Isolation of the thermobalance from pressure fluctuations in the chromatograph is the greatest difficulty associated with coupling TG–GC instruments (Figure 6.5). The decomposition of poly(ethylene–co-vinyl alcohol) recorded using TG–GC is presented in Figure 6.6.

6.1.4 TG–EGA REPORT

In addition to the items necessary to compile a complete TG report (Section 4.11), the following should be added when describing the results from a TG–EGA experiment:

- record of the evolved gas spectrum;
- description of how the gas components are identified;
- flow rate, total volume, design type and temperature of the interface between the TG and EGA instruments;

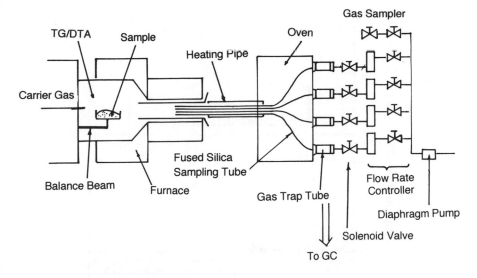

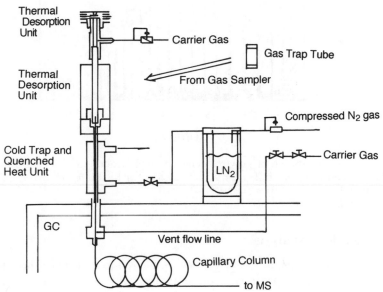

Figure 6.5. Schematic diagram of an integrated TG–DTA–GC apparatus (courtesy of Seiko Instruments and GL Sciences)

- type of EGA instrument, indicating location of the thermocouple used to determine the temperature of the evolved gases during analysis;
- delay between the evolution and analysis of gas;
- relationship between the signal amplitude and concentration of evolved gases.

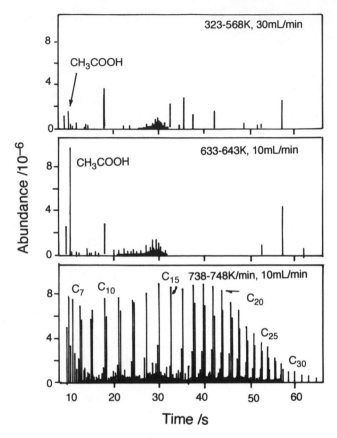

Figure 6.6. Decomposition of poly(ethylene–co-vinyl alcohol) as observed using GC. The simultaneously recorded TG curve is presented in Figure 6.2 (courtesy of Seiko Instruments)

6.2 Mechanical Analysis

Various forms of mechanical analysis are employed to determine the effect of thermal and chemical processing on polymers with a view to achieving a desired performance, or as a form of quality control. The two principal classes of mechanical analysis are thermomechanical analysis (TMA) and dynamic mechanical analysis (DMA). In TMA the deformation of a material under constant load (or constant strain) is recorded as a function of temperature or time. A sinusoidally varying stress is applied to the sample in DMA, producing an oscillating strain which lags behind the applied stress by a phase angle δ. The magnitude of the phase difference between the applied stress and the strain is a function of the structure of the material. In both methods the sample is subjected to a controlled temperature programme and controlled atmospheric conditions. Recently the distinction between TMA and DMA has become less clear as many modern TMA instruments apply an oscillating load to the sample.

6.2.1 THERMOMECHANICAL ANALYSIS (TMA)

TMA can be used to measure the deformation characteristics of solid polymers, films, fibres, thin films, coatings, viscous fluids and gels. Selection of the most appropriate load and deformation mode is important, and instruments are equipped with a number of attachments to optimize the experimental conditions (Table 6.1). A TMA apparatus which employs a balance beam mechanism in compression mode is shown in Figure 6.7. TMA curves are plotted with deformation on the vertical axis against temperature or time on the horizontal axis. Temperature calibration should be carried out unde experimental conditions identical with those for the proposed experiment. A disc (approximate thickness 0.1 mm) of material of well characterized melting point (Appendix 2.2) is placed in the TMA apparatus. Penetration of the reference material occurs on melting, giving rise to a large deformation signal, and the change in the shape of the TMA curve is used as a temperature calibration point (Figure 6.8A). A multi-point temperature calibration can be achieved in one run using a selection of standard materials in the sandwich configuration shown in Figure 6.8B. The

Table 6.1. TMA probes and deformation modes for specific applications

Sample	Parameter	Probe/deformation mode
Solid polymer	Linear expansion coefficient Glass transition temperature Softening temperature Melting temperature Creep, compliance	Compression/Expansion Flexure
Film, fibre	Young's modulus Glass transition temperature Softening temperature Creep, cure Cross-link density Thin films, coatings	Uniaxial extension/Shrinkage
	Young's modulus Glass transition temperature Softening temperature Creep, cure Cross-link density Hardness	Penetration
Viscous fluids, gels	Viscosity Gelation Gel–Sol transition Cure, elastic modulus	Shear Needle penetration

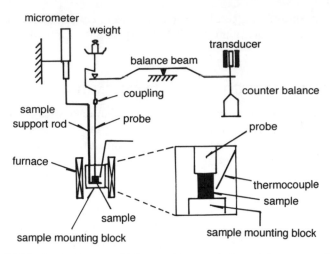

Figure 6.7. TMA instrument, which employs a balance beam mechanism, in compression mode (courtesy of Ulvac Sinku-Riko)

drawback of this method is that the standard samples can only be used once. The thermocouple which is used to record the sample temperature is rarely placed in contact with the sample, but is placed as close as possible to the sample. The sample-to-thermocouple distance should be maintained constant for all samples to minimize the effect of the atmospheric conditions in the sample chamber on the recorded sample temperature.

The probe displacement is calibrated using a micrometer or standard gauges whose thickness is precisely known. The applied load is calibrated using standard masses. On completion of the calibration procedures the instrument should be run under the proposed experimental conditions without the sample and the TMA curve recorded. This curve can be used later to correct for artifacts in the data which originate in the instrument.

The sample should be homogeneous, and where possible the upper and lower surfaces should be parallel and smooth. The samples used in TMA are relatively large and a heating (or cooling) rate of 1–5 K/min is recommended. Normally the chamber is maintained under dry N_2 at a flow rate of 10–50 ml/min. The mass of the selected probe should be taken into consideration when estimating the load applied to the sample.

TMA is used to determine the linear thermal expansion coefficient (α) of polymers, defined as

$$\alpha = \frac{dL}{dT} \frac{1}{L_0} \tag{6.1}$$

where L_0 is the original length of the sample and dL/dT is the slope of the TMA curve. The calculated value of α is temperature dependent

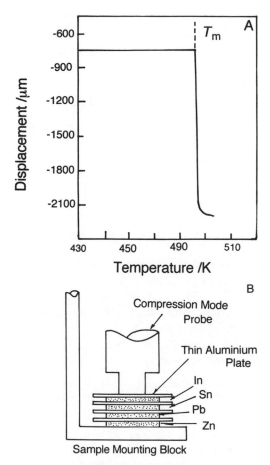

Figure 6.8. (A) TMA temperature calibration using tin as the standard reference material. (B) Sandwich configuration used to achieve a multi-point temperature calibration

(Figure 6.9). The glass transition temperature, T_g, of a sample can also be measured using TMA. T_g is the temperature at which an amorphous or semi-crystalline polymer is transformed from a rubbery viscous state to a brittle glass-like state. The measured value of T_g depends on the experimental conditions and the deformation mode employed. When measured by thermal expansion, T_g is the temperature at which the sample exhibits a significant change in its thermal expansion coefficient, under the given experimental conditions (Figure 6.10). Often it is easier to determine T_g from the derivative TMA curve. The value of T_g and/or α measured from the first experimental run may be significantly different from that of subsequent runs, as both of these parameters are dependent on the thermal history of the sample. The difference between the first and subsequent runs can reveal a great deal about the previous thermal history

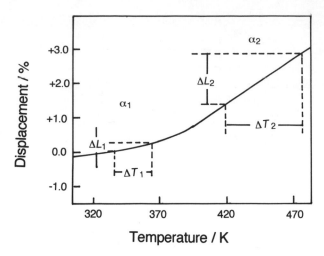

Figure 6.9. Determination of the linear thermal expansion coefficient (α) from a TMA curve

of the sample. The softening temperature is the temperature at which a material has a specific deformation, for a given set of experimental conditions. Although the softening temperature and T_g are related they are not equal, and a clear distinction should be made between them.

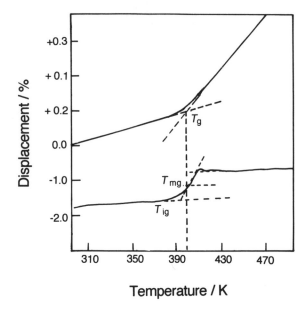

Figure 6.10. Determination of the glass transition temperature (T_g) from a TMA curve and the corresponding derivative TMA curve

Many polymers are viscoelastic and recover elastically following deformation. Figure 6.11 shows a schematic stress–strain curve where a tensile force is applied at a uniform rate to a viscoelastic sample at a constant temperature. The shape and characteristic parameters of the stress-strain curve are strongly influenced by the temperature and the sample processing conditions.

6.2.2 DYNAMIC MECHANICAL ANALYSIS (DMA)

In DMA the sample is clamped into a frame and the applied sinusoidally varying stress of frequency ω can be represented as

$$\sigma(t) = \sigma_0 \sin(\omega t + \delta) \qquad (6.2)$$

where σ_0 is the maximum stress amplitude and the stress proceeds the strain by a phase angle δ. The strain is given by

$$\gamma(t) = \gamma_0 \sin(\omega t) \qquad (6.3)$$

where γ_0 is the maximum strain amplitude. These quantities are related by

$$\sigma(t) = E^*(\omega)\gamma(t) \qquad (6.4)$$

where $E^*(\omega)$ is the dynamic modulus and

$$E^*(\omega) = E'(\omega) + iE''(\omega) \qquad (6.5)$$

$E'(\omega)$ and $E''(\omega)$ are the dynamic storage modulus and the dynamic loss modulus, respectively. For a viscoelastic polymer E' characterizes the ability of the polymer to store energy (elastic behaviour), while E'' reveals the

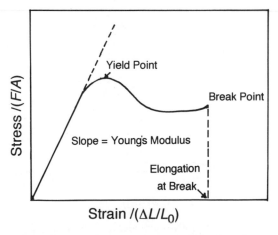

Figure 6.11. Schematic stress-strain curve for a viscoelastic polymer. The tensile force is applied at a uniform rate

tendancy of the material to dissipate energy (viscous behaviour). The phase angle is calculated from

$$\tan \delta = E''/E' \tag{6.6}$$

Normally E', E'' and $\tan \delta$ are plotted against temperature or time (Figure 6.12). DMA can be applied to a wide range of materials using the different clamping configurations and deformation modes (Table 6.2). Hard samples or samples with a glazed surface use clamps with sharp teeth to hold the sample firmly in place during deformation. Soft materials and films use clamps which are flat to avoid penetration or tearing. When operating in shear mode flat faced clamps, or clamps with a small nipple to retain the material, can be used. The head of the instrument can be damaged if the sample becomes loose during an experiment. Proper clamping is also necessary to avoid resonance effects. Computer-controlled DMA instruments allow the deforming force and oscillating frequency to be selected, and to be scanned automatically through a range of values, in the course of the experiment. DMA is a sensitive method to measure T_g of polymers. Side-chain or main-chain motion in specific regions of the polymer and local mode relaxations which cannot be monitored by DSC can be observed using DMA. From the variation in the temperature of the $\tan \delta$ peak of a DMA curve as a function of frequency a transition map can be compiled

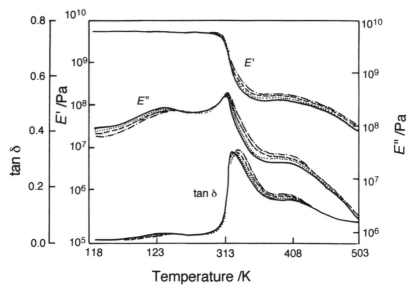

Figure 6.12. DMA curves of poly(vinyl alcohol) showing E', E'' and $\tan \delta$ as a function of temperature over a range of frequencies: ———, 0.5; ·····, 1.0; - - - -, 5.0; —·—·—, 10 Hz

Table 6.2.   DMA probes and deformation modes for specific applications

Sample	Parameter	Clamp/deformation mode
Solid polymer	Dynamic modulus	Flexure
	Glass transition temperature	Tension
	Melting temperature	Torsion
	Cross-link density	Compression
	Relaxation behaviour	
	Crystallinity, cure	Shear
Film, fibre, coatings	Dynamic modulus	Flexure
	Glass transition temperature	Tension
	Creep, cure, compliance	Shear
	Relaxation behaviour	
Viscous fluids, gels	Viscosity	Shear
	Gelation	
	Gel–Sol transition	
	Cure, dynamic modulus	

(Figure 6.13). If the locus of the transition map is a straight line, an activation energy for the phenomena responsible for the tan δ peak can be estimated using the Arrhenius relationship. When the locus is curved the Williams–Landel–Ferry (WLF) equation can be used to characterize the process. The calibration procedures and sample preparation methods are similar to those used in TMA.

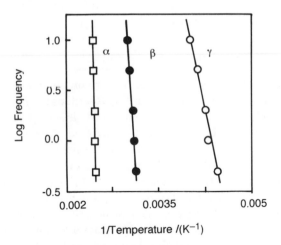

Figure 6.13.   Transition map of poly(vinyl alcohol) compiled using the DMA data presented in Figure 6.12. An activation energy for the α (motion in crystalline regions), β (glass transition) and γ (local mode relaxation) transitions can be calculated using the Arrhenius relation

6.2.3 TMA AND DMA REPORTS

The following items should be included along with the recorded TMA or DMA curves when presenting the results:

- sample identification and preconditioning;
- method of sample preparation, including dimensions and orientation (if applicable);
- type of TMA or DMA instrument used;
- deformation mode;
- shape and dimensions of probe (TMA);
- size and type of clamps, and frame (DMA);
- temperature range, heating/cooling rate, isothermal conditions;
- atmosphere, flow rate;
- description of temperature, displacement and load (force) calibration;
- exact location and type of sample thermocouple;

6.3 Dilatometry

Formerly dilatometry was commonly used to measure sample volume as a function of temperature. Glass capillary dilatometers were designed and built by individual researchers using mercury as the filling medium. Mercury is no longer used in volumetric experiments. Dilatometry is not as widely practised as before, in part because an alternative filling agent has not been found, and has been largely supplanted by TMA. Instead of the sample volume the linear expansion coefficient is measured using TMA (Section 6.2.1). However, the volume expansion coefficient cannot be estimated from TMA data since Poisson's constant is not 1.0 for many polymers.

6.3.1 DILATOMETER ASSEMBLY

Where a precise volumetric mesurement is required, a dilatometer can be constructed using the following procedure, whose steps are illustrated in Figure 6.14. A glass capillary 60–80 cm in length, whose inner tube diameter is 1 mm with an outer tube diameter of 5–7 mm, is selected. A glass tube 15-20 cm in length with a diameter of 15–20 mm and a wall thickness of less than 1 mm is connected to both ends of the capillary (step I). Another glass tube with the same dimensions is connected at an angle of 35–45°. This tube will serve as the mercury reservoir (step II). The sample (1–2 g) is inserted into the glass tube, followed by a glass rod of length 2–3 cm which fits the inner diameter of the glass tube and acts as a spacer (step III). The glass tube containing the sample is sealed using a gas burner and the glass capillary bent as shown in steps IV and V. The reservoir is filled with a precisely known amount of mercury. The dialtometer is connected to a vacuum line via a glass stopcock and evacuated (step VI). After evacuation, the stopcock is closed and the dilatometer discon-

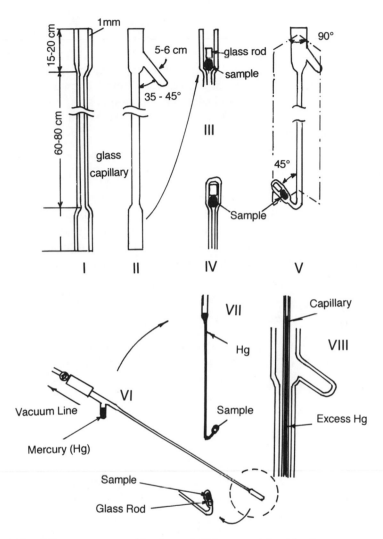

Figure 6.14. Dilatometer assembly. Steps I–VII are explained in the text

nected from the evacuation line. Holding the dilatometer in both hands, the dilatometer is rotated so that the mercury simultaneously fills the sample cell and capillary (step VII). A long glass capillary (60–80 cm) is prepared by stretching a glass tube using a gas burner. The outer diameter should be less than the inner diameter of the dilatometer's capillary tube. By inserting the newly made capillary into the dilatometer's capillary to approximately 5 cm higher than the sample in the dilatometer, an excess amount of mercury will fill the inserted glass capillary. The inserted capillary containing the excess mercury is removed and the excess mercury is transferred from the capillary into a weighing vessel so that the amount of mercury can be determined. The dilatometer containing the sample is placed in an oven and heated at a programmed rate. The height of the mercury in the glass capillary of the dilatometer is measured as a function of temperature. By this method, the volume expansion coefficient of the sample can be calculated if the sample mass and its density at room temperature are known, since the mass and the expansion coefficient of mercury and the diameter of the dilatometer capillary are known.

6.3.2 DEFINITION OF EXPANSION COEFFICIENTS

Three separate definitions of the thermal expansion coefficient are currently employed. When presenting data, or comparing a measured value with tabulated values, it is necessary to state clearly which definition is being used. If a solid sample is heated from T_1 to T_2 the length of the sample changes from L_1 to L_2 (Figure 6.15) and the linear expansion, α, at T_1 can be expressed as

$$\alpha = (1/L_1)[(L_2 - L_1)/(T_2 - T_1)] \tag{6.7}$$

$$\alpha = (1/L_1)(\Delta L/\Delta T) \tag{6.8}$$

When computers were not widely available, the above definition of α was not practical, since L_1 must be frequently measured during the heating process. A more convenient definition was used:

$$\alpha_1 = (1/L_0)[(L_2 - L_1)/(T_2 - T_1)] \tag{6.9}$$

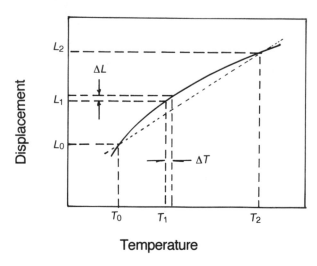

Figure 6.15. Various definitions of the linear expansion coefficient are currently employed using the parameters illustrated in this figure

where L_o is the length of the sample at 293 K. The International Standards Organisation uses this definition of α. Alternatively, α can be defined as

$$\alpha_0 = (1/L_0)\,[(L_2 - L_0)/(T_2 - T_0)] \qquad (6.10)$$

where T_o is 296 K or ambient temperature. The thermal expansion coefficient defined by equation 6.10 is used in many data tables. Since there are three definitions of the linear expansion coefficient there are three corresponding definitions of the expansion ratio, ε:

$$\varepsilon = (1/L_1)(L_2 - L_1) \qquad (6.11)$$

$$\varepsilon_1 = (1/L_0)(L_2 - L_1) \qquad (6.12)$$

$$\varepsilon_0 = (1/L_0)(L_2 - L_0) \qquad (6.13)$$

and three definitions of the volume expansion coefficient, β:

$$\beta = (1/V_1)\,[(V_2 - V_1)/(T_2 - T_1)] \qquad (6.14)$$

$$\beta_1 = (1/V_0)\,[(V_2 - V_1)/(T_2 - T_1)] \qquad (6.15)$$

$$\beta_0 = (1/V_0)\,[(V_2 - V_1)/(T_2 - T_0)] \qquad (6.16)$$

6.4 Thermomicroscopy

Thermomicroscopy is the characterization of a sample by optical methods while the sample is subjected to a controlled temperature programme, and can be used in conjunction with other TA techniques to record subtle changes in the sample structure. Solid-phase transformations, melting, crystallization, liquid crystallization and gel to liquid crystal transitions can be readily monitored by thermomicroscopy. In addition, decomposition, surface oxidation, swelling, shrinking, surface melting, cracking, bubbling and changes in colour and texture can be followed using thermomicroscopy with a sensitivity that is often greater than that of standard TA techniques. The principal modes of observation by thermomicroscopy are by reflected and by transmitted light.

6.4.1 OBSERVATION BY REFLECTED LIGHT

Alterations in surface structure alone rarely envolve large enough enthalpy fluxes to be detected by DSC, but do induce large changes in the reflected light intensity (RLI) from the surface. Although confined to the study of surfaces reflected light thermomicroscopy can be used with both opaque and transparent materials. The light source may be either a filament lamp (or a laser) and a photocell measures the changes in RLI as a function of temperature or time.

Simultaneous DSC–RLI apparatus have been constructed (Figure 6.16) where the sample is placed in an open DSC sample vessel. The sample should be as thin as possible to avoid thermal gradients between the surface and bulk of the material. Increased sample baseline curvature and a small reduction in DSC sensitivity are experienced under the open sample vessel conditions. Surface and interface effects can be probed by this method and the results used to determine their influence on the reaction kinetics of the sample.

6.4.2 OBSERVATION BY TRANSMITTED LIGHT

Measurements of the transmitted light intensity (TLI) can be more easily correlated with DSC results as this method records the effect of transformations occurring in the sample bulk on the transmitted light. This method is confined to transparent materials which are placed between glass slides for observation (Figure 6.17). The angle of rotation of transmitted polarized light is determined by the sample structure, and this method is widely used to study the nucleation and growth kinetics and the high order structure of liquid crystals. Simultaneous DSC–TLI instruments are commercially available (Figure 6.18), but due to design constraints the DSC sensitivity is lower than in the case of DSC–RLI.

6.5 Simultaneous DSC–X-Ray Analysis

Simultaneous DSC–X-ray analysis is a very powerful and versatile method for following changes in the morphology and structure of a wide range of materials under controlled temperature conditions. Instruments, based on those

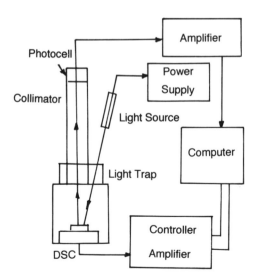

Figure 6.16.   Schematic diagram of a simultaneous DSC–RLI apparatus

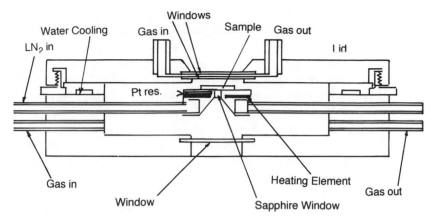

Figure 6.17. Microscope stage for TLI measurements (courtesy of Japan High-Tec)

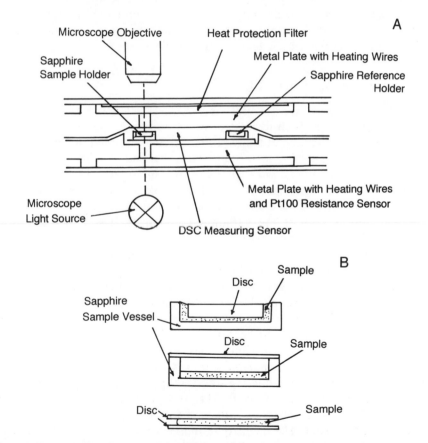

Figure 6.18. (A) Simultaneous DSC–TLI apparatus. (B) Sapphire sample holders (by permission of Mettler-Toledo)

developed for DSC–TLI (Section 6.4.2), are available for simultaneous DSC–small-angle X-ray scattering, wide-angle X-ray diffraction and synchrotron orbital radiation analysis (SAXS, $0.25° \leqslant 2\theta \leqslant 10°$; WAXD, $5° \leqslant 2\theta \leqslant 70°$; and SOR, $0.05° \leqslant 2\theta \leqslant 0.5°$). Given the broad angular range of these X-ray techniques, structural features ranging in size from 0.1 to 500 nm can be investigated. Owing to the high X-ray flux in SOR experiments, time-resolved X-ray analysis is possible. However, a large radiation flux can induce radiation damage in the form of main-chain, side-chain and cross-link scission. Where the transition temperature measured by X-ray analysis is consistently lower than that recorded using the DSC, for all scanning rates, the likelihood of radiation damage is high.

The sample vessel for simultaneous DSC–X-ray analysis must be made from materials of high transparency to X-rays and low diffuse scattering coefficient with few Bragg reflections, while at the same time possessing good thermal conductivity and exhibiting no phase changes in the temperature region of interest. Sample vessels made from aluminium, graphite and boron nitride are used. Data are plotted in the form of integrated scattering profile intensity and/or the DSC curve against temperature or time.

6.6 Thermoluminescence (TL)

Thermoluminescence (TL) measures the variation in intensity of luminescence of a sample which has been irradiated by UV radiation, X-rays, γ-rays or an electron beam as a function of temperature. Electrons excited by the impinging radiation become trapped in metastable states at liquid nitrogen temperatures. These electrons recombine with cations during subsequent heating owing to the enhancement of molecular motion in the sample. Luminescence is observed as energy is liberated by the electrons reverting to their ground state following recombination. A plot of the variation in intensity of luminescence with temperature is called a glow curve.

A block diagram of a TL instrument, which is composed of a light-proof box with a heating block to which the sample and a thermocouple are attached, is presented in Figure 6.19. An aluminium window allows the sample to be irradiated before heating. The chamber can be evacuated or purged with an inert gas. The TL sensor is a high sensitivity photomultiplier with a dark current $\leqslant 0.2$ nA. Typically the heating rate is 5 K/min and temperature calibration is carried out using low molecular mass, high purity n-alkanes. The wavelength of the TL from polymers ranges from 300 to 700 nm, but the intensity is generally weak rendering spectroscopic analysis difficult. An interference filter can be used to filter the TL at a preselected wavelength aiding analysis. The sample (1–50 mg) is attached to the heating block using silver electroconductive paint. The intensity of luminescence is low at high temperatures owing to recombination and therefore a large amount of sample should be used to improve the resolution in high temperature experiments.

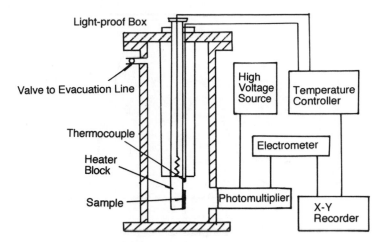

Figure 6.19. Schematic diagram of a thermoluminescence apparatus (courtesy of T. Hashimoto

TL characterizes the relaxation processes of electrons trapped in metastable states. Assuming that recombination is a first-order process, an activation energy for the liberation of the electrons can be calculated from the glow curve, using the Arrhenius relation. Under these assumptions the variation in intensity of TL with temperature is described by

$$I(T) = n_0 S \exp[-(E/kT) - (S/\beta) \int \exp(-E/kT)dT]   \qquad (6.17)$$

where n_0 (mol) is the initial concentration of electrons, S (s^{-1}) is a frequency factor, β (K/min) is the heating rate and E (J/mol) is the activation energy. E can be calculated from a plot of $\ln I$ versus $1/T$ using the slope of the low-temperature side of the glow peak and neglecting the integral term of equation 6.17. It is difficult to apply this simple analysis to the complex glow curves routinely recorded for polymers. Instead, peak-shape analysis is performed, varying E and S so that the calculated glow curve coincides with the experimental curve. The glow curve recorded for a polyacrylonitrile film is presented in Figure 6.20. By using curve-fitting software, the glow curve can be resolved into two distinct peaks. The low-temperature peak corresponds to relaxations occurring in the semi-crystalline regions of the film and the high-temperature peak is attributed to relaxations of the amorphous regions.

The following items should included when presenting the results of an TL measurement:

- type of TL apparatus;
- sample preparation including dimensions and method of attachment;
- temperature range and heating rate;
- irradiation source and total dose.

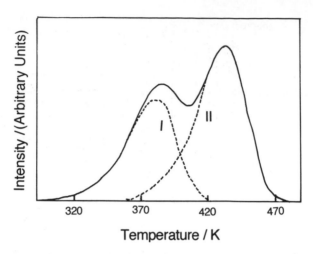

Figure 6.20. Glow curve of polyacrylonitrile. The low- and high-temperature peaks are attributed to relaxations in the crystalline regions and melting of the crystalline regions, respectively

6.7 Alternating Current Calorimetry (ACC)

Alternating current calorimetry (ACC) measures the alternating temperature change produced in a sample by an alternating heating current, from which the heat capacity of the material can be estimated. Assuming that heat does not leak from the sample during heating and a constant current amplitude and frequency, the amplitude of the alternating temperature of the sample is given by

$$T_{ac} = (Q/i\omega C_p)e^{i\omega t} \tag{6.18}$$

where C_p is the heat capacity of the sample, $Q\,e^{i\omega t}$ is the heat flux and ω the angular frequency. A block diagram of an AC calorimeter is presented in Figure 6.21. The output of a white light source is modulated with a variable-frequency beam chopper so that a square wave is produced which illuminates one surface of the sample. The fluctuating alternating temperature is measured on the other surface using a thermocouple. Figure 6.22 shows the variation in T_{ac} as a function of time for materials of large and small heat capacity. Owing to recent improvements in the design of lock-in amplifiers which can operate at low frequencies, ACC can be applied to a broad range of materials, including polymers.

The operating temperature range is typically 100–1000 K using samples of area 30–50 mm^2 and thickness 0.01–0.3 mm. The temperature resolution is ± 0.0025 K for $T < 770$ K and ± 0.025 K for $T > 770$ K. The sample holder is purged with a dry inert gas. Alumel–chromel or chromel–constantan thermocouples of 0.002 mm diameter are placed in a paper frame and soldered to

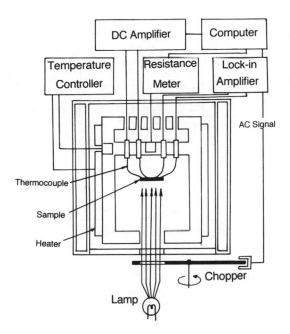

Figure 6.21. Schematic diagram of an alternating current calorimeter

metal samples to measure T_{ac}. Polymers are dissolved in an organic solvent and the solution is spread on a thin metal support (stainless-steel film), which is then placed in an evacuated oven to dry the sample. The thermocouple is fixed to the metal support in this case. For polymers, the accuracy of the C_p measurement is $\pm 2\%$ in absolute value. Fine graphite powder in sol form is sprayed on the illuminated surface of the polymer to ensure complete absorption of the impinging light (Figure 6.23).

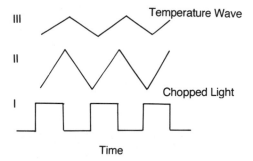

Figure 6.22. (I) In ACC the sample is illuminated by a white light source whose output is modulated to produce a square wave. Schematic alternating temperature, as a function of time, for materials of (II) small and (III) large heat capacity

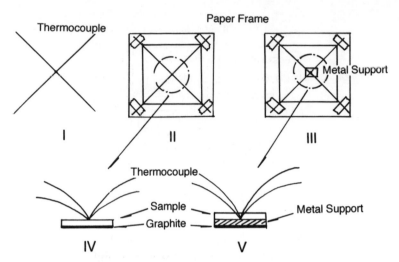

Figure 6.23. Preparation of sample for ACC. Steps I, II and IV are used for samples to which the thermocouple can be directly attached. Polymers are deposited on a thin metal support to which the thermocouple is fixed (steps I, II, III and V)

An ACC curve for polystyrene is shown in Figure 6.24, where it can be seen that the variation in C_p as a function of temperature can be continuously monitored even in the region of a phase change.

The following items should included when presenting the results of an ACC measurement:

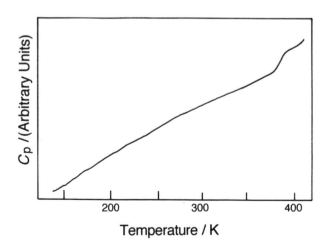

Figure 6.24. Heat capacity of polystyrene as a function of temperature recorded using ACC

- type of ACC apparatus;
- purge gas and pressure;
- method of sample preparation including dimensions;
- type of metal support including dimensions;
- thermocouple type and method of attachment;
- temperature range and heating rate;
- illumination frequency.

6.8 Thermal Diffusivity (TD) Measurement by Temperature Wave Method

Thermal diffusivity can be measured under non-steady-state conditions using laser-flash and photoacoustic methods. Because the size of the sample used in these techniques is relatively large, the time to attain thermal equilibrium is long, rendering these methods unsuitable for measuring the thermal diffusivity of polymers over a broad temperature range at a programmed heating rate. A non-steady-state method based on the principle of the temperature wave has been developed to measure the thermal diffusivity of organic materials in the solid and liquid state, which can be employed under scanning temperature conditions. An alternative Joule heat applied to the front surface of a sample generates a periodic heat flow as the heat diffuses to the rear surface of the sample. The thermal diffusivity is estimated from the phase difference between the input and output alternating voltages.

A block diagram of a thermal diffusivity apparatus and a sample holder are presented in Figure 6.25. A polymer film (approximate thickness 50 µm) is sandwiched between glass plates, the inner surfaces of which have been sputtered with gold. One sputtered layer is used as a heating plate and the other as a resistance sensor. Thick strips of gold are sputtered on top of the original sputtered layer so that the electrical connections can be made. The edges of the glass plates are sealed with an epoxy resin. For liquid samples a spacer is placed between the plates. The glass assembly is placed on a copper hot-stage whose temperature is controlled by a temperature programmer. An alternating electric current is applied using a function synthesizer to the heating plate, and the Joule heat induces a temperature wave which propagates through the sample. Temperature fluctuations in the rear surface of the sample are detected as a variation in the electrical resistance of the gold layer. The phase lag between the signal applied to the heating plate by the function synthesizer and the voltage measured at the rear surface is measured using a lock-in amplifier.

The periodic temperature fluctuation can be described using the thermal diffusion equation assuming one-dimensional heat diffusion in the sample, good thermal contact between the sample and glass plate via the sputtered gold layer and that the temperature of the glass remains constant. Under these assumptions, the phase lag $\Delta\theta$ is given by

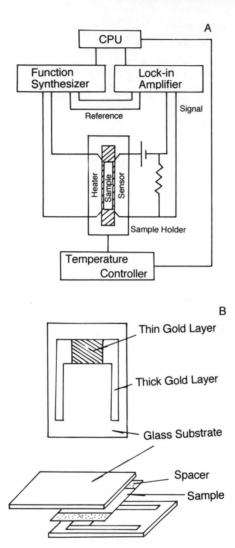

Figure 6.25 (A).   Schematic diagram of a thermal diffusivity apparatus. (B) Preparation of sample for TD measurement (courtesy of T. Hashimoto)

$$\Delta\theta = d(\omega/2\alpha)^{\frac{1}{2}} + \pi/4 \qquad (6.19)$$

where α is the thermal diffusivity and d the sample thickness.

The thermal diffusivity of an n-alkane as a function of temperature and the corresponding DSC curve are presented in Figure 6.26. The peak on the low-temperature side of the DSC curve is attributed to a crystal–crystal transition and on the high-temperature side to melting. The value of α decreases at each transition temperature.

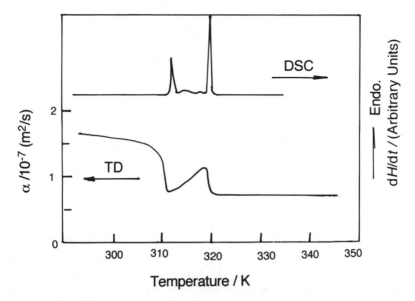

Figure 6.26. Thermal diffusivity of *n*-alkane as a function of temperature and the corresponding DSC curve. The low-temperature feature is due to a crystal–crystal transition and the high-temperature feature to melting

The following items should included when presenting the results of a TD measurement:

- type of TD apparatus;
- method of sample preparation including dimensions;
- temperature range and heating rate;
- frequency of applied Joule heat.

6.9 Thermally Stimulated Current (TSC)

Analysis by thermally stimulated current (TSC) measures the relaxation processes occurring in a sample which has been polarized at a temperature greater than the temperature where molecular motion in the sample is enhanced, and subsequently quenched so that the high mobility state is frozen. On heating the sample, depolarization of polymer electrets (dipoles, trapped electrons, mobile ions) occurs and the oriented dipoles frozen in the quenched sample relax to a state of thermal equilibrium. This relaxation process is observed as a depolarization current, which is typically of the order of picoamperes, and is referred to as the thermally stimulated current.

A TSC apparatus is shown schematically in Figure 6.27. In practice, the sample is sandwiched between parallel condenser-type electrodes and heated to the depolarization temperature, T_p. An electric field (E_p), which is applied

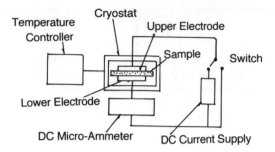

Figure 6.27. Schematic diagram of a thermally stimulated current apparatus (courtesy of H. Shimizu)

across the sample for a time t_p, polarizes the sample. The sample is then quenched at a constant cooling rate to T_0, maintaining the applied electric field. After quenching, the electric field is extinguished and the sample heated to obtain a TSC. This procedure is illustrated in Figure 6.28, I.

A Debye-type single relaxation process is normally assumed when analysing the shape of the TSC curve, and an activation energy can be estimated using the Arrhenius relationship from the initial gradient of the TSC curve. However, it is unreasonable to assume that a single relaxation process is responsible for the complex TSC curves recorded for polymers. In order to deconvolute these curves into individual relaxation processes where the Debye

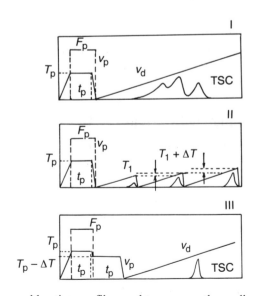

Figure 6.28. Poling and heating profiles used to measure thermally stimulated current. (I) Standard measurement, (II) partial heating method or peak cleaning method; (III) thermal sampling method

and Arrhenius relationships are more applicable, two approaches are used. The first is called the partial heating method or peak cleaning method (Figure 6.28, II). Following quenching and the removal of the applied electric field, the TSC is measured as the sample is heated at a programmed rate to T_1. The sample is requenched from T_1 to T_0 and the TSC measured while heating to $T_1 + \Delta T$ at the same heating rate as before. Typically $\Delta T \approx 10\,\mathrm{K}$. The sample is once again quenched and heated to $T_1 + 2\Delta T$. This procedure is repeated until the TSC is no longer observed. In the thermal sampling method (Figure 6.28, III) the sample is polarized in the high mobility state and subsequently quenched to $T_p - \Delta T$. The electric field is removed and relaxation occurs at $T_p - \Delta T$ for t_i. On heating to T_p a TSC is observed for those electrets which did not relax when the sample was maintained at $T_p - \Delta T$ for t_i. At T_p the electric field is again applied and the sample quenched to $T_p - 2\Delta T$ from where the measuring cycle repeated. In this case $\Delta T \approx 5\,\mathrm{K}$.

The following items should included when presenting the results of a TSC measurement:

- type of TSC apparatus;
- method of sample preparation including dimensions;
- electric field, poling temperature and poling time;
- temperature following quenching;
- temperature range and heating rate;
- deconvolution method.

APPENDIX 1
GLOSSARY OF TA TERMS

Adiabatic calorimeter: Instrument for measuring the absolute heat capacity of a substance under quasi-equilibrium conditions.

Alternating current calorimeter: Instrument for measuring the alternating temperature change produced in a substance by an alternating heating current.

Alternating current calorimetry: Branch of thermal analysis, where the alternating temperature change produced by an alternating heating current is used to investigate the nature of a substance.

Automatic sample supplier: Robot arm for routine loading and removal of samples from thermal analysis instruments.

Balance: Instrument for measuring mass.

Baseline: See *Instrument baseline* and *Sample baseline*.

Bending mode: Configuration of TMA (or DMA) instrument, where a sample is fixed at both ends and a constant (or oscillating) stress is applied.

Cooling rate: Rate of temperature decrease in response to a temperature programme.

Creep curve: Graphical representation of the time-dependent strain of solid materials caused by constant applied stress.

Crucible: Vessel used to hold sample, particularly in thermobalances.

Crystallization: Formation of crystalline substances from solutions, melts or the glassy state.

Curie temperature: Temperature of transition from ferromagnetism to paramagnetism, or from a ferromagnetic phase to a paramagnetic phase.

Derivative thermogravimetric (DTG) curve: Graphical representation of the data collected by a thermobalance, where the rate of change of mass with respect to temperature (or time) is plotted as a function of temperature (scanning mode) or time (isothermal mode).

Derivatogram: General term for derivative TA curve.

Differential scanning calorimeter: Instrument for measuring the differential energy supplied between a sample and reference to maintain a minimal temperature difference between the sample and reference in response to a temperature programme.

Differential scanning calorimetry (DSC): Branch of thermal analysis where the differential energy supplied between a sample and reference to maintain a minimum temperature difference between the sample and reference in response to a temperature programme is used to investigate the nature of the sample.

Differential scanning calorimetry curve: Graphical representation of the data collected by a differential scanning calorimeter, where the differential energy supplied is plotted as a function of temperature (scanning mode) or time (isothermal mode).

Differential thermal analyser: Instrument for measuring the difference temperature between a sample and reference in response to a temperature programme. Also known as classical differential thermal analyser.

Differential thermal analysis (DTA): Branch of thermal analysis where the difference temperature between a sample and reference in response to a temperature programme is used to investigate the nature of the sample.

Differential thermal analysis curve: Graphical representation of data collected by a differential thermal analyser, where the difference temperature is plotted as a function of temperature (scanning mode) or time (isothermal mode).

Dilatometer: Instrument for measuring the thermal expansion and dilation of liquids and solids.

Dynamic mechanical analyser: Instrument for measuring the behaviour of a sample subjected to an oscillating stress in response to a temperature programme.

Dynamic mechanical analysis (DMA): Branch of thermal analysis where the behaviour of a sample subjected to an oscillating stress in response to a temperature programme is used to investigate the nature of the sample.

Dynamic mechanical analysis curve: Graphical representation of the data collected by a dynamic mechanical analyser, where the dynamic loss modulus, dynamic storage modulus and tan δ are plotted as a function of temperature (scanning mode) or time (isothermal mode).

Endotherm: Deviation from the sample baseline of a DSC (or DTA) curve indicating energy absorption by the sample relative to a reference.

Enthalpy: Sum of the internal energy of a system plus the product of the system volume multiplied by the ambient pressure.

Exotherm: Deviation from the sample baseline of a DSC (or DTA) curve indicating energy release by the sample relative to a reference.

Extrapolated sample baseline: Extension of the sample baseline of a DSC (or DTA) curve into the region of a phase change, used to calculate the characteristic temperatures and enthalpy change associated with the change of phase.

Fusion: See *Melting*.

Glass transition: Change of state of an amorphous or semi-crystalline polymer from a rubbery (or viscous) state to a glassy state. The glass transition is not a thermodynamic first- or second-order phase transition. It is a relaxation phenomenon which is characterized by a general enhancement of molecular motion in the polymer at the glass transition temperature.

Glass transition temperature: Temperature of transition of an amorphous or semi-crystalline polymer from a rubbery (or viscous) state to a glassy state.

Heat capacity: Quantity of heat required to raise the temperature of a system by 1 K at constant pressure (or constant volume).

Heat conductivity: See *Thermal conductivity*.

Heat-flux type DSC: Commercial name for quantitative DTA.

Heating rate: Rate of temperature increase in response to a temperature programme.

Instrument baseline: DSC (or DTA) curve recorded in the scanning mode when there is no sample or reference present.

Isothermal mode: Operating mode of TA instruments, where the response of the sample is monitored as a function of time at a fixed temperature.

Linear thermal expansion: Expansion of sample in one direction in response to a temperature programme.

Melting: Change of state of a substance from a solid phase to a liquid phase. Also known as fusion.

Melting temperature: Temperature of transition from a solid phase to a liquid phase.

Modulated DSC: Variation of DSC (or quantitative DTA) where a sinusoidal perturbation is applied to the temperature programme resulting in a cyclic modulation of the heat flow and temperature signals. The precise nature of the modulation is at present unknown. Also known as oscillated DSC.

Onset temperature: Transition temperature defined as the intersection between the tangent to the maximum rising slope of a DSC (or DTA) peak and the extrapolated sample baseline.

Oscillated DSC: See *Modulated DSC*.

Peak: General term for an endothermic or exothermic deviation from the sample baseline.

Phase: Homogeneous portion of a system (liquid, gas, solid) with distinct boundaries which can be distinguished from other phases of the system.

Phase diagram: Graphical representation of the phase structure of a system as a function of an experimental parameter (pressure, temperature, composition, etc.)

Phase transiton enthalpy: Enthalpy change of a system due to a change of phase.

Phase transition temperature: Temperature of transition from one phase of a system to another phase.

Power compensation-type differential scanning calorimeter: Instrument for measuring the differential electric power supplied between a sample and reference to maintain a minimal temperature difference between the sample and reference, in response to a temperature programme.

Purge gas: Inert gas which replaces the atmosphere in the vicinity of a sample to standardize the experimental conditions.

Quantitative differential thermal analyser: Instrument for measuring the difference temperature between a sample and reference in response to a temperature programme. Knowing the heat capacity of the heat-sensitive plate as a function of temperature, this instrument can be used to estimate the enthalpy change associated with a change of phase in the sample. Also known as heat-flux differential scanning calorimeter.

Reference: Substance whose instantaneous temperature and heat capacity are continuously compared with that of the sample over the entire temperature range of a DSC (or DTA) measurement. The reference is generally inert over the temperature range of the measurement.

Sample baseline: Linear portion of a DSC (or DTA) curve, recorded in the presence of a sample and reference, outside the transition region.

Sample holder: Device used to house the sample in a TA instrument. The sample is placed in a sample vessel in DSC, which is inserted into the sample holder.

Sample holder assembly: Module of DSC (or DTA) instrument consisting of the sample and reference holders and the associated mechanical supports, electrical connections and heat sources.

Sample vessel: Receptacle for sample in DSC (or DTA) which can be made from a variety of materials, including aluminium, gold and silver.

Standard reference material: High-purity material exhibiting a well characterized phase change which is used to calibrate a TA instrument.

Stress-relaxation curve: Graphical representation of the time-dependent stress of solid materials caused by constant strain.

Stress–strain curve: A graphical representation of the relationship between the stress applied to a sample and the strain (or deformation) that results.

Tensile mode: Configuration of TMA (or DMA) instrument where a sample is subjected to a constant (or oscillating) longitudinal stress.

Thermal analysis (TA): Class of analytical methods where the nature of a sample is investigated in response to a temperature programme. Includes DMA, DSC, DTA, TG and TMA. Also known as thermoanalysis.

Thermal conductivity: Heat flow across a surface per unit area per unit time, divided by the rate of change of temperature with distance in a direction perpendicular to the surface. Also known as heat conductivity.

Thermal diffusivity: Quantity of heat passing normally through unit area per unit time divided by the product of specific heat, density and temperature gradient.

Thermally stimulated current (TSC): Electric current observed following the depolarization of a sample through heating. The sample is initially poled in an electric field at a temperature greater than the glass transition or melting temperature and subsequently quenched.

Thermobalance: Instrument for measuring the mass change of a sample in response to a temperature programme.

Thermocouple: A device composed of two dissimilar conductors joined at both ends, where a voltage is developed in response to a temperature difference between the junctions. Once calibrated a thermocouple can be used to measure the temperature of a system to a high degree of accuracy.

Thermogravimetry (TG): Branch of thermal analysis where the mass change of a sample in response to a temperature programme is used to investigate the nature of the sample. Also known as thermogravimetric analysis (TGA).

Thermogravimetry curve: Graphical representation of data collected by a thermobalance, where the mass change is plotted as a function of temperature (scanning mode) or time (isothermal mode).

Thermoluminescence (TL): Branch of thermal analysis where the variation in intentsity of luminescence of a sample which has been irradiated by UV radiation, an electron beam, X-rays or γ-rays, in response to a temperature programme, is used to investigate the nature of the sample.

Thermomechanical analyser: Instrument for measuring the behaviour of a sample subjected to a constant stress in response to a temperature programme.

Thermomechanical analysis (TMA): Branch of thermal analysis where the deformation of a sample subjected to a constant stress in response to a temperature programme is used to investigate the nature of the sample.

Thermomechanical analysis curve: Graphical representation of data collected by a thermomechanical analyser where the deformation of the sample is plotted as a function of temperature (scanning mode) or time (isothermal mode).

APPENDIX 2
STANDARD REFERENCE
MATERIALS

A2.1 Temperature and Enthalpy of Fusion of Recommended Standard Reference Materials
(courtesy of T. Matsuo)

Reference material	T_{im}/K	$\Delta H_{fus}/J/g$
Indium (In)	429.78	28.5 ± 0.2
Tin (Sn)	505.12	59.7 [1], 60.6 [2], 59.6 [3], 56.57 ± 0.10 [4]
Lead (Pb)	600.65	23.2 ± 0.5
Zinc (Zn)	692.73	111.18 ± 0.44 [5]
Aluminium (Al)	933.45	398 [1], 388 [3], 399 [6]
Silver (Ag)	1235.08	107 [1], 105 [3], 112 [6]
Biphenyl	342.41	120.41 [7]

[1] Speros, D.M. and Woodhouse, R.L. *Journal of Physical Chemistry* **67**, 2164 (1963).
[2] Grønvold, F. *Revue de Chimie Minerale* **11**, 568 (1974).
[3] Kubaschewski, O. and Alcock, C.B.. *Metallurgical Thermodynamics*, 5th ed, Pergamon Press, Oxford, 1979.
[4] NIST (NBS) SRM 2220.
[5] NIST (NBS) SRM 2221.
[6] Kelley, K.K. *U.S. Bureau of Mines Bulletin* **584**, 1960.
[7] NIST (NBS) SRM 2222.

A2.2 Standard Reference Material Sets Sold by the ICTAC through NIST

Set No.	Reference material	T_{im} /K
GM 754	Polystyrene	378
GM 757	1,2-Dichloroethane	241
	Cyclohexane (transition)	190
	Cyclohexane (melting)	280
	Diphenyl ether	303
	o-Terphenyl	331
GM 758	Potassium nitrate	401
	Indium	430
	Tin	505
	Potassium perchlorate	573
	Silver sulphate	703
GM 759	Potassium perchlorate	573
	Silver sulphate	703
	Quartz	846
	Potassium sulphate	856
	Potassium chromate	938
GM 760	Quartz	846
	Potassium sulphate	856
	Potassium chromate	938
	Barium carbonate	1083
	Strontium carbonate	1198
GM 761 (TG)	Permanorm 3	532
	Nickel	626
	Mumetal	654
	Permanorm 5	727
	Trafoperm	1023

A2.3 Heat Capacity Data of Sapphire (α-Al$_2$O$_3$) as a Function of Temperature

Molar mass of sapphire: 101.9612 g/mol.

T/K	C_p/J/g K	T/K	C_p/J/g K	T/K	C_p/J/g K
100	0.1260	390	0.9295	680	1.1392
110	0.1602	400	0.9423	690	1.1430
120	0.1969	410	0.9544	700	1.1467
130	0.2350	420	0.9660	720	1.1537
140	0.2740	430	0.9770	740	1.1604
150	0.3133	440	0.9875	760	1.1667
160	0.3525	450	0.9975	780	1.1726
170	0.3913	460	1.0070	800	1.1782
180	0.4291	470	1.0160	820	1.1836
190	0.4659	480	1.0247	840	1.1887
200	0.5014	490	1.0330	860	1.1936
210	0.5355	500	1.0408	880	1.1984
220	0.5682	510	1.0484	900	1.2030
230	0.5994	520	1.0556	920	1.2074
240	0.6292	530	1.0626	940	1.2117
250	0.6576	540	1.0692	960	1.2158
260	0.6845	550	1.0756	980	1.2197
270	0.7101	560	1.0816	1000	1.2237
280	0.7342	570	1.0875	1020	1.2275
290	0.7571	580	1.0931	1040	1.2311
300	0.7788	590	1.0986	1060	1.2347
310	0.7994	600	1.1038	1080	1.2383
320	0.8188	610	1.1088	1100	1.2417
330	0.8372	620	1.1136	1120	1.2450
340	0.8548	630	1.1182	1140	1.2484
350	0.8713	640	1.1227	1160	1.2515
360	0.8871	650	1.1270	1180	1.2546
370	0.9020	660	1.1313	1200	1.2578
380	0.9161	670	1.1353		

C_p calculated from the following: $C_p/\text{J/g K} = C(0) + C(1)x + \ldots + C(10)x^{10}$, $100 \leqslant T/\text{K} \leqslant 1200$ and $x = (T/\text{K} - 650)/550$.

$C(0) = 1.12705$	$C(4) = -0.23778$	$C(8) = -0.47824$
$C(1) = 0.23260$	$C(5) = -0.10023$	$C(9) = -0.37623$
$C(2) = -0.21704$	$C(6) = 0.15393$	$C(10) = 0.34407$
$C(3) = 0.26410$	$C(7) = 0.54579$	

APPENDIX 3
PHYSICAL CONSTANTS AND CONVERSION TABLES

A3.1 Table of Physical Constants

Quantity	Symbol	Value
Permeability of vacuum	μ_0	$4\pi \times 10^{-7}$ H/m
Velocity of light	c	299 792 458 m/s
Dielectric constant of vacuum	$\varepsilon_0(\mu_0 c^2)^{-1}$	$8.854\,187\,816 \times 10^{-12}$ F/m
Fine-structure constant	$\alpha = \mu_0 c e^2/2h$	$7.297\,353\,08(33) \times 10^{-3}$
	α^{-1}	$137.035\,989\,5(61)$
Electronic charge	e	$1.602\,177\,33(49) \times 10^{-19}$ C
Planck's constant	h	$6.626\,075\,5(40) \times 10^{-34}$ Js
	$\hbar = h/2p$	$1.054\,572\,66(63) \times 10^{-34}$ Js
Avogadro's number	L, N_A	$6.022\,136\,7(36) \times 10^{23}$ mol^{-1}
Atomic mass unit	amu	$1.660\,540\,2(10) \times 10^{-27}$ kg
Electron rest mass	m_e	$9.109\,389\,7(54) \times 10^{-31}$ kg
Proton rest mass	m_p	$1.672\,623\,1(10) \times 10^{-27}$ kg
Neutron rest mass	m_n	$1.674\,928\,6(10) \times 10^{-27}$ kg
Faraday's constant	$F = Le$	$9.648\,530\,9(29) \times 10^4$ C/mol
Rydberg constant for infinite mass	$R_\infty = \mu_0{}^2 m_e e^4 c^3/8h^3$	$1.097\,373\,153\,4(13) \times 10^7$ m^{-1}
Hartree energy	$E\mathrm{a} - 2hcR_\infty$	$4.359\,748\,2(26) \times 10^{-18}$ J
First Bohr radius	$a_0 = \alpha/4\pi R_\infty$	$5.291\,772\,49(24) \times 10^{-11}$ m
Bohr magneton	$\mu_B = eh/2m_e$	$9.274\,015\,4(31) \times 10^{-24}$ J/T
Nuclear magneton	$\mu_N = eh/2m_p$	$5.050\,786\,6(17) \times 10^{-27}$ J/T
Magnetic moment of electron	μ_e	$9.284\,770\,1(31) \times 10^{-24}$ J/T
Landé g-factor for free electron	$g_e = 2\mu_e/\mu_B$	$2.002\,319\,304\,386(20)$
Proton gyromagnetic ratio	γ_p	$2.675\,221\,28(81) \times 10^8$ s^{-1}T^{-1}
Gas constant	R	$8.314\,510(70)$ J/K mol
0°C in Kelvin	T_0	273.15 K
	RT_0	$2.271\,108(19) \times 10^3$ J/mol
Atmospheric pressure	p_0	101 325 Pa
Molar volume of ideal gas	$V_0 = RT_0/p_0$	$2.241\,410(19) \times 10^{-2}$ m^3/mol
Boltzmann's constant	$k = R/L$	$1.380\,658(12) \times 10^{-23}$ J/K
Acceleration due to gravity	g	$9.806\,65$ m/s^2

A3.2 Energy Conversion Table

J	cal	BTU	kWh	atm	kg m
1	0.239006	0.947831×10^{-3}	2.777778×10^{-7}	9.86896×10^{-3}	0.101972
4.184	1	3.96573×10^{-3}	1.162222×10^{-6}	4.12929×10^{-2}	0.426649
1.055040×10^{3}	2.52161×10^{2}	1	2.930667×10^{-4}	10.41244	1.07584×10^{2}
3.6×10^{6}	8.60421×10^{5}	3.41219×10^{3}	1	3.55292×10^{4}	3.67098×10^{5}
1.01325×10^{2}	24.2173	9.60390×10^{-2}	2.814583×10^{-5}	1	10.33223
9.80665	2.34385	9.29505×10^{-3}	2.724069×10^{-6}	9.67841×10^{-2}	1

A3.3 Molar Energy Conversion Table

J/mol	erg/mol	cal/mol	eV/mol	cm^{-1}	K
1	$1.660\,566 \times 10^{-17}$	$0.239\,006$	$1.036\,435 \times 10^{-5}$	$8.359\,348 \times 10^{-2}$	$0.120\,273\,1$
$6.022\,045 \times 10^{16}$	1	$1.493\,03 \times 10^{16}$	$6.241\,461 \times 10^{11}$	$5.034\,037 \times 10^{15}$	$0.724\,290 \times 10^{16}$
4.184	$6.947\,806 \times 10^{-17}$	1	$4.336\,444 \times 10^{-5}$	$0.349\,755$	$0.503\,2227$
$9.648\,455 \times 10^{4}$	$1.602\,189 \times 10^{-12}$	$2.306\,036 \times 10^{4}$	1	$8.065\,479 \times 10^{3}$	$1.160\,450 \times 10^{4}$
$11.962\,655$	$1.986\,477 \times 10^{-16}$	$2.859\,143$	$1.239\,852 \times 10^{-4}$	1	$1.438\,786$
$8.314\,41$	$1.380\,663 \times 10^{-16}$	$1.987\,192$	$8.617\,347 \times 10^{-3}$	$0.695\,0304$	1

A3.4 Pressure Conversion Table

Pa, N/m^2	Torr, mmHg	bar	kg/cm^2	psi	atm
1	7.50062×10^{-3}	10^{-5}	1.01972×10^{-5}	1.45038×10^{-4}	9.86923×10^{-6}
133.322	1	1.3322×10^{-3}	1.35951×10^{-3}	1.93368×10^{-2}	1.31579×10^{-3}
10^5	750.062	1	1.01972	14.5038	0.986932
9.80665×10^4	735.559	0.980665	1	14.2233	0.967841
6.89476×10^3	51.7149	6.89476×10^{-2}	7.03070×10^{-2}	1	6.80460×10^{-2}
1.01325×10^5	760	1.01325	1.03323	14.6959	1

A3.5 Thermal conductivity Conversion Table

J/s m K	kcal/m h °C	cal/cm s °C
1	0.860 421	$2.390\,06 \times 10^{-3}$
1.162 22	1	$2.777\,78 \times 10^{-3}$
418.4	360	1

A3.6 Temperature Conversion Equations

$T/°C = T/K - 273.15 = (T/°F - 32) \times 5/9$
$T/°F = (T/K) \times 9/5 - 459.67 = (T/°C) \times 9/5 + 32$
$T/K = T/°C + 273.15 = (T/°F + 459.67) \times 5/9$

CHEMICAL FORMULA INDEX

n-Alkane

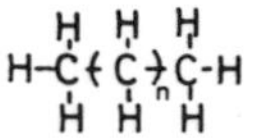

$$H-\overset{\overset{\displaystyle H}{|}}{\underset{\underset{\displaystyle H}{|}}{C}}(\overset{\overset{\displaystyle H}{|}}{\underset{\underset{\displaystyle H}{|}}{C}})_n\overset{\overset{\displaystyle H}{|}}{\underset{\underset{\displaystyle H}{|}}{C}}-H$$

α-Alumina $\alpha\text{-Al}_2\text{O}_3$

Amide group $-\text{CONH}_2$

Amino acid $\text{NH}_2-\text{M}-\text{COOH}$
R = aliphatic radical

Aromatic carboxylic acid

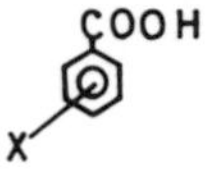

Aromatic polyester   Polyesters with phenyl group in the main chain

Benzene

Benzoic acid

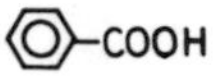

Butadiene rubber   Viscoelastic synthetic polymer made from butadiene. *See* Polybutadiene

Cellulose

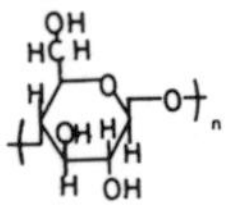

Cotton   Natural fibre obtained from plants of the genus *Gossypium*.
See Cellulose

DNA   Deoxyribonucleic acid. A polymer composed of deoxyribonucleotide repeating units. Most DNA molecules form double-stranded, antiparallel helices

Diphosphorus pentaoxide P_2O_5

Epoxy resin   A polyether resin formed by the polymerization of bisphenol A and epichlorohydrin

Hexacosane   *See* *n*-Alkane, $n = 14$

Hydroxyl group $-\text{OH}$

Natural rubber	A natural high polymer with elastic properties and, after after vulcanization, elastic recovery

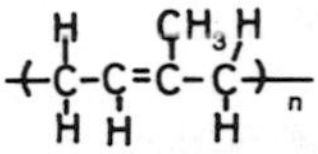

$$\left[\begin{array}{c} H \quad CH_3 \; H \\ -C-C=C-C- \\ H \; H \quad\quad H \end{array}\right]_n$$

Nichrome	Ni–Cr alloy

Polyacrylonitrile (PAN)

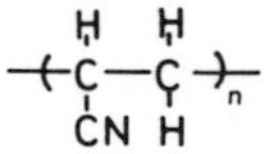

$$\left[\begin{array}{c} H \quad H \\ -C-C- \\ CN \; H \end{array}\right]_n$$

Polyamino acid	Polymer composed of amino acid repeating units

Polybutadiene

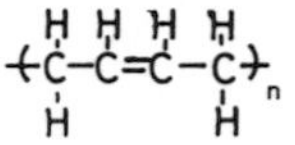

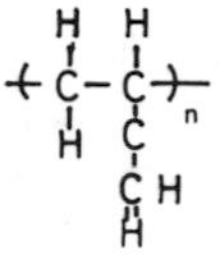

$$\left[\begin{array}{c} H \; H \; H \; H \\ -C-C=C-C- \\ H \quad\quad H \end{array}\right]_n \qquad \left[\begin{array}{c} H \; H \\ -C-C- \\ H \; C \\ \;\; C \; H \\ \;\; H \end{array}\right]_n$$

Polycarbonate (PC)

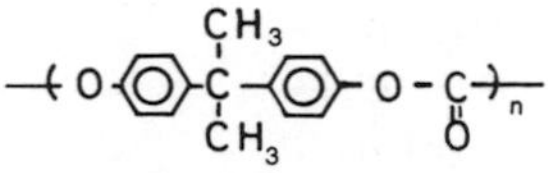

$$\left[O-\bigcirc-\underset{CH_3}{\overset{CH_3}{C}}-\bigcirc-O-\underset{O}{C} \right]_n$$

Polyethylene (PE)

$$\left[\begin{array}{c} H \; H \\ -C-C- \\ H \; H \end{array}\right]_n$$

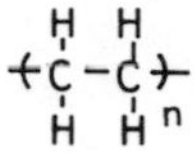

Poly(ethylene–co–vinyl alcohol) (EVA)	Ethylene–vinyl alcohol random copolymer

Poly(ethylene terephthalate) (PET)

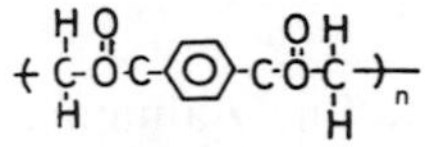

$$\left[\underset{H}{\overset{H \; O}{C}}-O-C-\bigcirc-C-O-\underset{H}{\overset{O \; H}{C}} \right]_n$$

Poly(4-hydroxystyrene) (PHS)

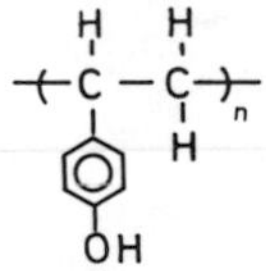

$$\left[\begin{array}{c} H \quad H \\ -C-C- \\ \;\;\; H \\ \bigcirc \\ OH \end{array}\right]_n$$

Poly(methyl methacrylate) (PMMA)

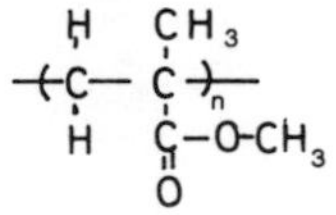

$$\left[\begin{array}{c} H \quad CH_3 \\ -C-C- \\ H \quad C-O-CH_3 \\ \quad\;\; O \end{array}\right]_n$$

Poly(oxymethylene) (POM)

$$\left[\begin{array}{c} H \\ -C-O- \\ H \end{array}\right]_n$$

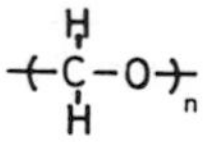

Poly(phenylene oxide (PPO) $\left[\bigcirc-O\right]_n$

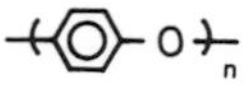

Polypropylene (PP)

$$\left[\begin{array}{c} H \; H \\ -C-C- \\ CH_3 \; H \end{array}\right]_n$$

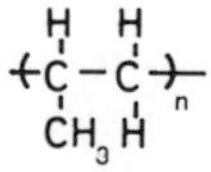

Polysaccharide	A carbohydrate composed of monosaccharide repeating units

Polystyrene (PSt)

$$\begin{array}{cc} H & H \\ | & | \\ -\!\!\!\left(\!\!\begin{array}{c} \\ C \\ | \\ H \end{array}\!\!-\!\!\begin{array}{c} \\ C \\ | \\ \phi \end{array}\!\!\right)\!\!\!-_{n} \end{array}$$

Poly(thio-1,4-phenylene
phenylphosphonyl-
1,4-phenylenethio-4,4'-
biphenylene)

Poly(tetrafluoroethylene)
(PTFE)

$$-\!\!\!\left(\!\!\begin{array}{cc} F & F \\ | & | \\ C & -C \\ | & | \\ F & F \end{array}\!\!\right)\!\!\!-_{n}$$

Poly(vinyl acetate) (PVAc)

$$-\!\!\!\left(\!\!\begin{array}{cc} H & H \\ | & | \\ C & -C \\ | & | \\ H & O\!-\!C\!-\!CH_3 \\ & \| \\ & O \end{array}\!\!\right)\!\!\!-_{n}$$

Poly(vinyl alcohol) (PVA)

$$-\!\!\!\left(\!\!\begin{array}{cc} H & H \\ | & | \\ C & -C \\ | & | \\ OH & H \end{array}\!\!\right)\!\!\!-_{n}$$

Poly(vinylidene fluoride)
(PVDF)

$$-\!\!\!\left(\!\!\begin{array}{cc} Cl & F \\ | & | \\ C & -C \\ | & | \\ F & F \end{array}\!\!\right)\!\!\!-_{n}$$

Protein   Polymer composed of various α-amino acids
                            joined by peptide linkages

Quartz SiO_2

Sapphire   Crystalline α-alumina

Schizophyllan

Silicon carbide   SiC

Toluene

Xanthan

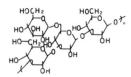

Subject Index